THE UNDERGROUND GUIDE TO LASER PRINTERS

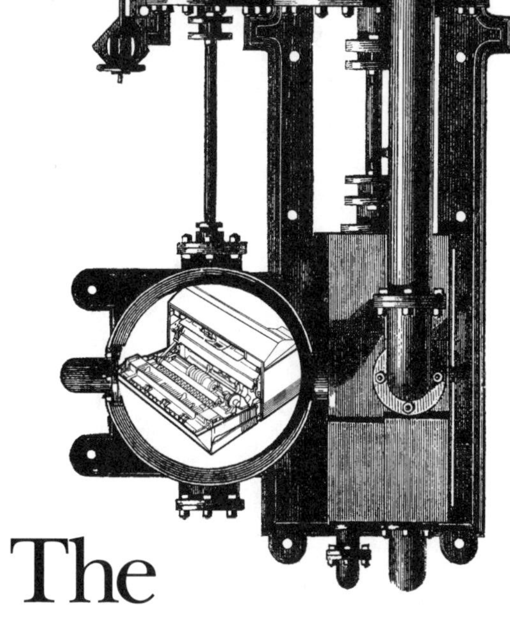

The Underground Guide to Laser Printers

by
the Editors of
Flash Magazine

OPEN HOUSE

PEACHPIT PRESS

THE UNDERGROUND GUIDE TO LASER PRINTERS
by the editors of *Flash Magazine*

PEACHPIT PRESS, INC.
2414 Sixth St.
Berkeley, CA 94710
(800) 283-9444
(510) 548-4393
(510) 548-5991 (fax)

Cover design by Studio Silicon

Cover illustration by Ark Stein

Interior design and production by Olav Martin Kvern

ISBN 1-56609-045-8

0 9 8 7 6 5 4 3 2 1

Printed and bound in the United States of America

Contents

Preface

The Birth of this Book

When we began putting out *The Flash,* a small four-page newsletter about laser printers for our customers, in early 1989, we had no idea that it would grow to be larger than what was then our main business: remanufacturing laser printer toner cartridges with specialty toners. Four years later, *Flash Magazine* is available on newsstands nationwide, has spawned two books, a CD-ROM, and has grown to a circulation many times that of our laser printer toner cartridge business.

Flash readers are the people who have driven the evolution by asking for more articles, insisting on paying for subscriptions (originally *The Flash* was free), and by their thousands of requests for a book like this, containing the best of the back issues of *The Flash.* The main reasons mentioned by readers for their love of *The Flash* are the "high meat-to-fat content," the readability of the articles, and the dash of Yankee ingenuity—the unique flavor of Vermont in each issue.

There has always been a high demand for back issues of *The Flash,* so we have always printed several thousand extra copies of each issue. But all good things come to an end, and so did the supply of back issues. As we began running out, readers suggested that we put together some sort of collection of reprints. We started taking the idea seriously after receiving more than a thousand such requests in less than a year.

This book is the result. In it you will find the best articles about laser printers from the first four and a half years of *The Flash.* Articles explaining how printers work. Articles about cleaning, maintaining, and even repairing your laser printer. Articles that help you get the best possible copy from your printer. Articles on wild ideas in laser printing such as colorful iron-on transfer toners, laser color foils, making rubber stamps, and more.

Most of the articles were group collaborations. At the head of each article you'll find the name of the primary author, as well as when the article originally appeared in *The Flash.* Scattered in with the articles you'll also find Flash Bulbs—little hints and tips on making the most of your laser printer.

Many people have contributed over the past four and a half years to create what went into this book—from the employees of BlackLightning who worked so hard to make each issue, to the outside contributors who have written some of the articles in recent issues; from the advertisers who have contributed financially to the production of *The Flash,* to the tens of thousands of readers who have given us feedback and asked the questions that prompted and guided *The Flash*'s development over the years.

A project like this has too many contributors to thank by name, but we would like to mention a few who have made outstanding contributions over the years. Thanks to Melissa Kalarchian, who helped me get the first issue out the door. Thanks to Holly Blumenthal, John Jeffries, Richard Alexander, Catherine Croft, Rebecca Mae, Wendy Hutchins, Dawn Marie Poland, Tom Durgin, Barbara Brooks, Dick Dermody, and Richard Hyde for the many long days, late nights, and hard weekends of research, writing, and proofing that it has taken to create each issue and this book. Thanks to Jill, Charlie, Scott, Shawn, Ray, Trav, and the rest of the kids from West Topsham who help with labeling and mailing out the tens of thousands of issues and a million other little things. Thanks to Richard Sanford, Jr. for his financial backing and faith. Thanks to Charlene, our West Topsham Postmaster, for her patience and assistance with all our mailings. And finally, thanks to Ted Nace and Steve Roth of Peachpit Press who helped us take this book on the final leg of its journey to bookstores and you, the reader. We hope you enjoy this, and perhaps you'll even want to subscribe to *Flash* Magazine. Write us with your comments. We love hearing from you!

Walter Vose Jeffries
Editor and Publisher
Flash Magazine

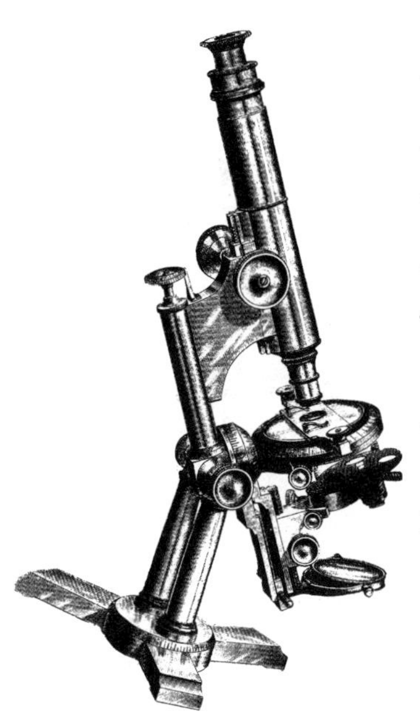

How Laser Printers Work

Toner, Transfer, Heat, and Pressure

By Holly Blumenthal of BlackLightning

From *The Flash*, Volume 2, Issue 2

As with artists who make their dancing and drawing look easy, the laser printer is a well-engineered piece of machinery that makes an intricate, complex process happen consistently and smoothly. Here, we will look at the inner workings of the popular H-P Series II laser printer. This machine is similar to most laser printers, and the general principles discussed are applicable to both laser printers and photocopiers.

The Paper Tray

Life begins at the paper tray. The top sheet in the tray is picked up by rollers located at the end of the paper tray, inside the printer (Figure 1). These rollers are shaped like a flat tire, and are covered with a textured plastic which helps grab the paper. From the paper tray, the sheet is passed through the machine by a series of rollers. This paper path passes under the cartridge to pick up the toner, and then through the fuser assembly, where heat and pressure melt the toner onto the page. Finally, the copy emerges from the printer and comes to rest on the output paper tray.

FIGURE 1
Printer operations

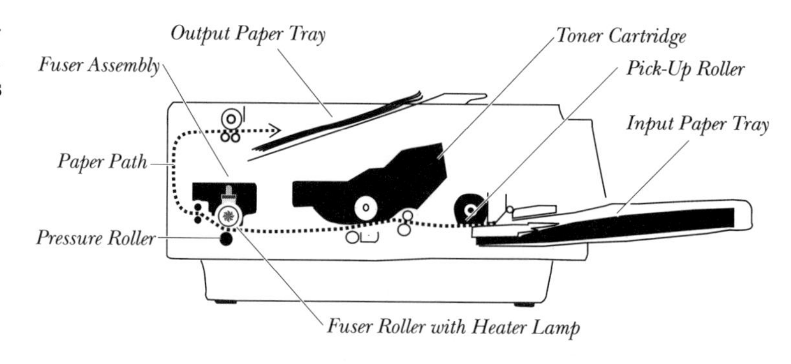

Feed Me

When not used regularly, the manual feed section of a laser printer may become dirty and cause smudges on the printout. To clean this area, simply feed several pages through the printer before use. Heavy paper works better.

Jammin' & Crammin'

Persistent paper jams in your laser printer or copier can be caused by several things: a foreign object in the paper path; dirty paper rollers; overheating, which causes internal rollers to expand and misalign (clean or replace the ozone filter); worn exit rollers, or worn lower fuser rollers. If you get a persistent paper jam, notice where it is occurring and look at the rollers nearby. They are likely culprits.

As the paper begins its journey, the cartridge drum is being prepared for the developing process. This means not only cleaning off any residual toner, but also clearing the cartridge drum surface of electrostatic charges. Electrostatic charges are a critical part of the laser printing process. The voltage and type of charges involved vary widely from machine to machine. Thus, the many different types of toner available. For simplicity of explanation, positive and negative signs are used in this article with no specific voltages. The cartridge drum is specifically designed to receive charges. The outer layers of the drum are made of an Organic Photo Conductive material. Hence the name: OPC drum. This photosensitive layer becomes electrically conductive

FIGURE 2
The photoelectric process

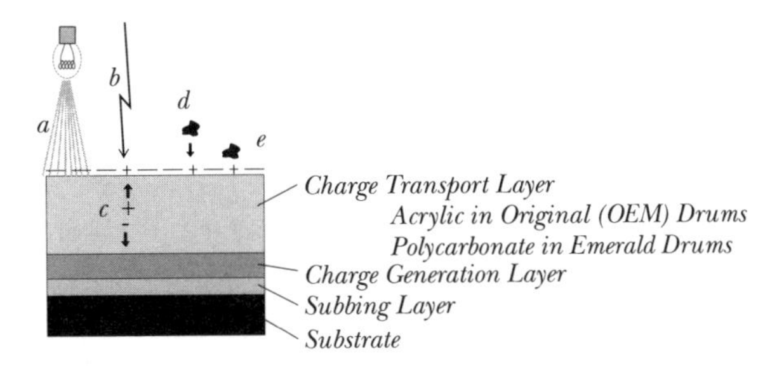

when exposed to light (Figure 2), much like the solar cells used in many calculators. Beneath this is an aluminum substrate.

The Eraser Lamp

Inside the printer is an eraser lamp, which shines through a slot in the top of the cartridge onto the OPC drum (Figure 3). This lamp clears the drum of any previous electrostatic charge, providing a "clean slate." The drum of the cartridge rotates constantly during the printing process, exposing the entire surface of the drum to the eraser lamp as well as to the rubber wiper blade, which sweeps excess toner from the drum into the waste reservoir.

This wiper blade eventually wears fine grooves into standard OPC drums, because they are made of a soft acrylic. The grooves then carry excess toner that shows up as black marks on the printed page. This is why the super-hard polycarbonate sur-

FIGURE 3
Inside an EP-S cartridge

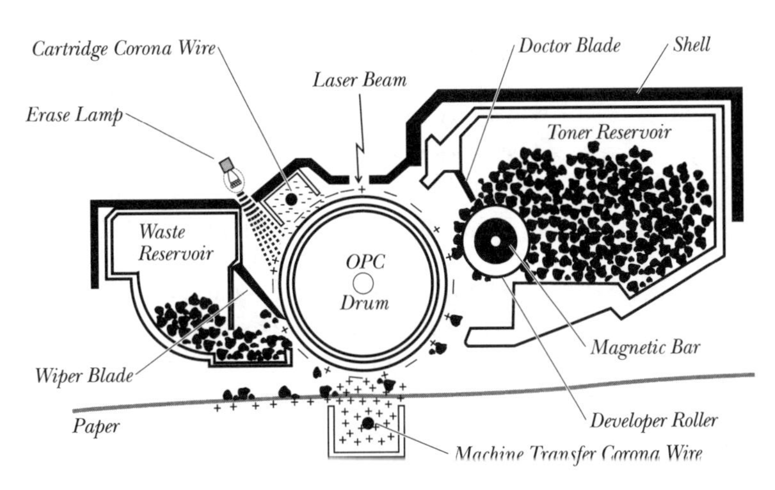

face of the Emerald OPC drum results in better print quality and a longer life.

The clean, discharged drum surface passes under the corona wire in the cartridge. This corona wire applies a uniform negative charge to the drum surface. It is something like painting a canvas with whitewash—a uniform, negative electrostatic charge. Dirt on the corona wire can prevent the charge from being uniform, and may result in black smears on the final output.

At this point, the drum is ready to receive the image. The printer's laser beam is reflected off a rotating, six-sided mirror, and directed down to the drum surface through a slot in the cartridge case. The laser printer translates the digital image from the computer into a sequence of on/off instructions for the laser beam. The laser can turn on and off at a rate of 30,000 times per second. Wherever the laser beam light is focused, the electrostatic charge on the drum surface becomes more positive. The whitewash was a negative charge; the new image is more positive. This is called the electrostatic image (Figure 4). The positive charges of this example will pick up the toner, as we will soon see.

Toner

The "ink" of the laser printer is actually a very fine powder called toner. The primary components of toner are plastic, sand, and rust. Toners have an electrostatic charge. Between the toner reservoir and the drum of the cartridge is the developer roller (Figure 3). The developer roller has a magnet inside it which attracts the already charged toner, and increases the strength of the toner's charge. The doctor blade, located just above the roller surface, is adjusted to allow the correct amount of toner to be passed onto the drum. This adjustment is referred to as "gapping," and it controls the flow and charging of toner.

The electrostatic image of our example is positively charged. The toner is negativly charged. Opposites attract; thus, toner is drawn to the drum, changing the electrostatic image into a picture that can be seen on the drum surface (Figure 4).

The paper is now passing under the cartridge and over the machine corona wire (Figure 3). The corona wire in the machine applies a strong positive charge to the paper. This pulls the negatively charged toner down off the drum and onto the sheet. The image can now be seen on the page. Don't sneeze; the powdery toner is only lying in place on the surface of the paper as it passes over the paper guide. Sometimes, when the machine jams, it is necessary to open the top and remove the partially printed paper. You can easily smear the image on one of these pages.

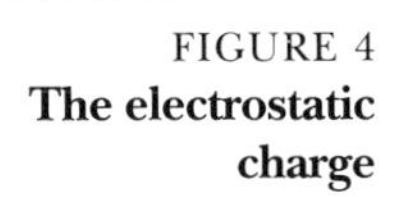
FIGURE 4
The electrostatic charge

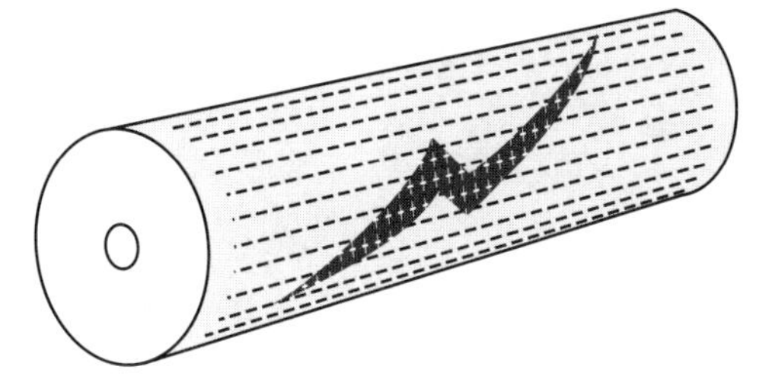

Fuser Rollers

Next, the paper passes between the two rollers of the fuser assembly, located beneath the fuzzy green cover that says "Warning: High Temperatures." The top roller of the fuser assembly has a high-intensity heat lamp in it which heats the roller to approximately 300° F. The lower roller is the pressure roller. The combination of heat and pressure melts the toner onto the sheet. The fuser wand has a specially oiled felt that lubricates the fuser rollers and cleans the top roller of paper dust and toner. The printed sheet emerges from the fuser assembly.

That is the completion of the paper's expedition. The gears, electronics, computer interpretation, mirrors, lasers, toner resins, heat, and pressure all combine to work as an intricate, synchronized whole. With this well-conducted orchestration of events, the image prints smoothly and easily.

Laser Printer Engines

Hundreds of laser printers are based on just a few laser engines

By John and Walter Jeffries

Despite the bewildering array of laser printers available, there are only a few companies that make the print engines that drive these printers. Most of the printer manufacturers (paradoxically called *original equipment manufacturers,* or *OEMs*) like Apple, H-P, QMS, Xante, LaserMaster, etc., buy the laser engine (paper-feed mechanism, laser controller, and toner system) from one of these companies. They then add the computer interface, page description language, and printer controller, wrap it all in a pretty case, and slap a label on it.

The upshot of this is that printers made by one OEM often use the same supplies as the printers from another OEM. This fact is useful for two reasons: First, the price of the supplies varies significantly depending on who is selling them and for which OEM they are sold—even when the supplies are identical. Second, the supplies for some of the print engines can be recycled by third-party cartridge remanufacturers. The determining factor is the design of the print engine, not whose name is on the printer.

One-, Two-, and Three-Part Consumables

There are three common approaches to the consumables in a laser printer. The all-in-one cartridges hold all the consumables in one, relatively expensive, easily installed unit that must be replaced each time the printer runs out of toner. It was Canon's use of this approach in their very popular print engines that led to the explosive growth of the cartridge recharging industry. When you run out of toner in a toner cartridge of this type, many of the components are still very serviceable.

The two-component approach separates the toner reservoir and the developer unit from the photosensitive drum. This way you only have to replace the less expensive toner/developer unit when it is out of toner. The more expensive photosensitive drum is replaced much less often. A few cartridge recyclers will work with these cartridges, though the economics are less compelling than those of the all-in-one cartridges.

The three-component approach makes it possible for the consumer to refill the toner reservoir in the printer with a specially made toner bottle. Thus you only replace the very cheap toner frequently, the more expensive drum occasionally, and the developer unit is almost never replaced. It is not cost-effective to try to refill these toner "cartridges"; the shipping costs alone will kill any savings.

Characteristics of Laser Engines

Below are the distinguishing characteristics of a variety of print engines, with a picture of the toner cartridge and a short list of the major OEM users of that engine. Use these descriptions to determine the type of print engine that you have. For a more complete list of printers by brand name, see the Machine Compatibility Chart later in this book.

A word is in order concerning the number of pages you can expect to get from a particular cartridge. You must first determine how much toner will be used on each page—the average coverage. A standard definition of the average coverage is 5 percent; this amounts to a short letter covering less than one third of the page with text, no letterhead and no graphics. A higher average coverage yields fewer pages and a lower average coverage yields more pages. For example, if you normally print full pages of text, the coverage would be 15 to 20 percent, and you will probably only get a third to a quarter as many pages as the manufacturer claims.

The choice of 5 percent was arrived at because originally the manufacturers thought that users would not print their letterhead at the same time as the letter. They were expected to put the letterhead into the printer and print just the text of the letter. With the advent of PostScript and high-quality toners, this number is rather low for many people, who now print the letterhead and the letter in the same pass, or do desktop publishing with full pages of text and graphics. The page count in the following chart is based on 5 percent coverage on letter-size paper, even for tabloid printers.

TABLE 1 **Laser printer cartridges**

Engine Make Cartridge Name Toner System	**What the Cartridge Looks Like**	**Distinguishing Characteristics**	**Pages per refill**	**Example printers**
Canon CX EP All-in-one		The cartridge has a very distinctive handle, like a briefcase. The entire top of the printer hinges up like a clam shell, and the cartridge slides into the side of the top half.	3,000	H-P LaserJet, LaserJet Plus, Apple LaserWriter, LaserWriter Plus
Canon SX EP-S All-in-one		The cartridge is shaped like a wedge with its tip cut off. There are two plastic doors on the top of the cartridge. The back half of the top of the printer hinges back, and the cartridge slides point first into the top.	4,500	H-P LaserJet II, IID, III, IIID, Apple LaserWriter II, IINT, IIf, IIg.
Canon NX IIIsi All-in-one		The cartridge is shaped like a wedge with its tip cut off. The laser and the erase openings are not covered on this cartridge. The back half of the top of the printer hinges back, and the cartridge slides point first into the top.	8,000	HP LaserJet IIIsi

TABLE 1 **Laser printer cartridges** (continued)

Engine Make **Cartridge Name** **Toner System**	**What the Cartridge Looks Like**	**Distinguishing Characteristics**	**Pages per refill**	**Example printers**
Canon LX EP-L All-in-one	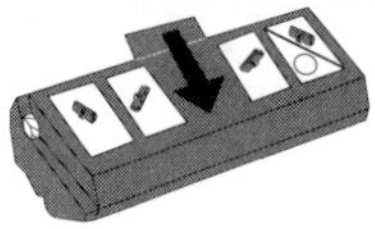	The cartridge is roughly box-shaped, with rounded corners. It has a 3½-inch-wide plastic tab in the middle used to pull it out of the printer. The front of the printer hinges down and forward, and the cartridge slides into the main part of the printer from the front.	3,000	HP LaserJet IIP, IIIP QMS 410, Apple Personal LaserWriter IINT
Canon EX EP-E All-in-one	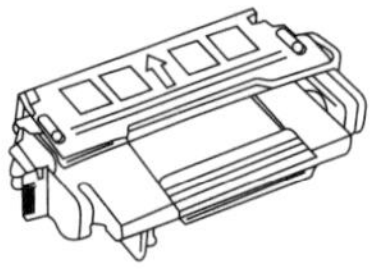	This print engine is one of the first true 600-dpi print engines from Canon. The back half of the top of the printer hinges back, and the cartridge slides point first into the top.	6,000	HP LaserJet IV and the Apple Laser-Writer Pro 600 and 630
Canon BX EP-B All-in-one	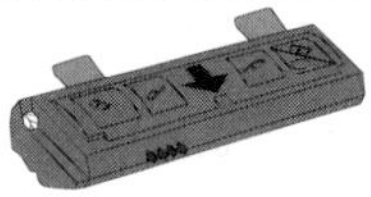	This print engine produces true 600-dpi output and can print on paper as large as 11 by 17 inches. The front of the printer folds down and the cartridge slides into the body of the printer from the front.	6,000	Xante 8100, NewGen TurboPS 660B, QMS 860
Minolta Imaging Cartridge 2 All-in-one	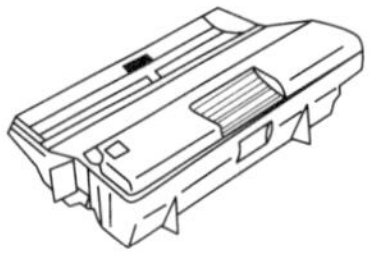	This cartridge is roughly rectangular and has many gears on the right side, visible from the bottom. Part of the top hinges up and the cartridge slides in to the top part.	8,000	NEC SilentWriter II 90, Acer Grayscale printer
IBM 4019 4019 All-in-one	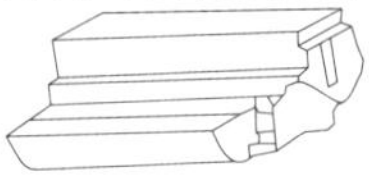	This cartridge looks like two steps, with a rounded edge under the "lower" step. Part of the top of the printer hinges up and the cartridge slides in, blunt end first.	5,600*	IBM 4019 printers
Fuji/Xerox EP All-in-one	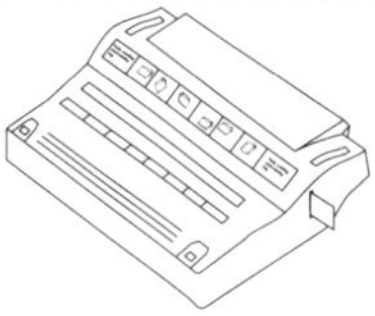	The cartridge is shaped like a wedge with its tip cut off. It prints on paper up to 11 by 17 inches. The back half of the top of the printer hinges back, and the cartridge slides point first into the top.	12,000	Dataproducts LZR 1560

*IBM claims that the IBM 4019 high-yield cartridge will print 10,000 pages. However they use 2.8% as the average page coverage. This works out to about 5,600 pages at the normal 5% coverage.

TABLE 1 **Laser printer cartridges** (continued)

Engine Make Cartridge Name Toner System	What the Cartridge Looks Like	Distinguishing Characteristics	Pages per refill	Example printers
Sharp 9500 Three-component		The toner cartridge is rectangular with a narrow spot in the middle. The top of the printer tips up, and the cartridge slides into the top.		Sharp JX 9500
Kyocera TK-4 Three-component		The toner cartridge is almost exactly rectangular. A door in the top of the printer opens to admit the cartridge.		Kyocera F-1010, F-2010, F-1000, F-1200
Ricoh 4080 R4080 Three-component		This toner cartridge is also almost exactly rectangular, with a lip all the way around the base of the cartridge.		Ricoh 4080, QMS 1500, DEC Scriptprinter
Ricoh 6000 R6000 Three-component		This cartridge has a special crank that is used to pull back the sealing strip. The printer opens like a clam shell, and the cartridge is placed in the bottom half.		Ricoh 6000

CX Maintenance

Care and Feeding of Your Series I Laser Printer

By Walter Vose Jeffries of BlackLightning
From *The Flash,* Volume 1, Issue 1

Laser printers have become an integral part of many computer users' lives. They typically perform flawlessly, printing crisp copy day after day, with nary a thought to maintenance or cleaning by their users. But a day without your laser printer, when you need to print that critical report, is like a day without sunshine.

Laser printers are low-maintenance machines, but they can get quite dirty from paper dust and spilled toner, as well as from dust in the environment. A clean machine will produce better output, remain trouble-free longer, and is relatively simple to clean. Let's take a look at cleaning the original Apple LaserWriter and LaserWriter Plus, H-P LaserJet Series 1, QMS KISS, and QMS PS-800, all of which are based on the Canon CX laser engine.

First, a quick introduction is in order. Figure 1 shows some of the key components of the LaserWriter. Start by opening the machine: gently but firmly pull up on the release lever, as indicated by the black arrow. Then let's jump into the most critical part of the cleaning process.

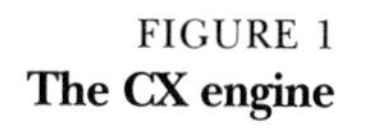

FIGURE 1
The CX engine

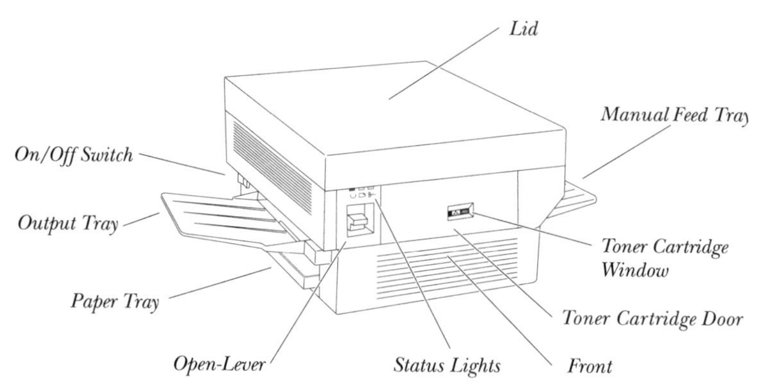

The CX Laser Printer

With the printer lid open, look in the upper part of the printer above where the cartridge handle is normally located. There are two filters here that often become clogged, resulting in the overheating of the machine. This can cause blurred letters and graphics on your printed copy. They can be cleaned gently with a vacuum or a cloth.

Find the silver-colored rectangular well next to the black plastic ridged paper path in the center of the machine. It is crisscrossed by very fine wires. These are the printer corona and transfer wires. Using a cotton swab moistened with alcohol, gently wipe each of the wires (Figure 2). The corona wire is the barely visible long wire under the short, diagonally crossed transfer wires. This should also be cleaned. Be gentle; they are fragile and expensive to replace.

Signs that the corona wire need cleaning are white patches or vertical streaks, like drips of paint, on your printed copy. In this case, you should also clean the corona wire in the cartridge

FIGURE 2
Cleaning the transfer wires

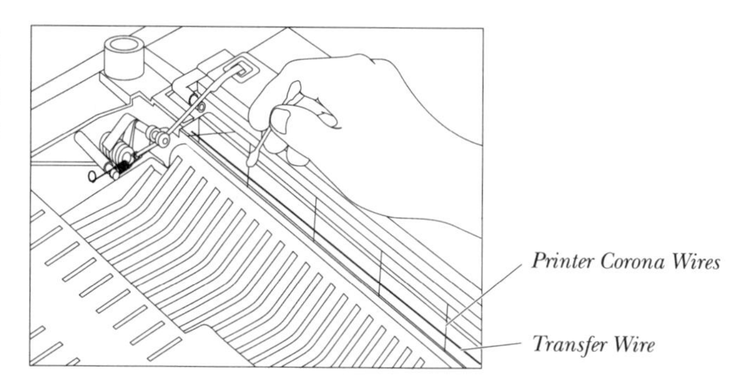

(more later). Caution: do not clean the printer corona wire with the cartridge corona wire cleaner. This will dirty the tool excessively, and reverse its intended usefulness. Once it is dirty, it will only make the cartridge corona wire dirtier. When used for its intended purpose, it does not get very dirty.

If a large amount of toner builds up in the well, a vacuum cleaner is the best way to get it out. Again, be careful with the fragile wires.

Wipe the transfer guide with a barely damp cloth. This is the bronze-colored plate you will see behind the transfer wire when you open the printer. Pay especially close attention to the space below the top plate.

Replacing the Fuser Wand

The fuser wand is the black plastic wand with the fuzzy green handle. The felt strip on the bottom cleans the fuser rollers that bind the toner to the paper with heat and pressure. Check the white felt of the fuser wand for buildup of residue. If it is excessively grimy, gently scrape the crust into the trash with the edge of a paper clip or similar firm edge. If this does not help, the wand needs to be replaced. The wand is vital for the proper operation of the laser printer, and should be replaced whenever the cartridge is changed. Extra wands may be purchased from dealers and toner cartridge remanufacturers.

The Separation Belt

The separation belt (Figure 3) is a 3-inch-long clear plastic strip that guides the paper and prevents jamming. It is located on the opposite side of the machine from the release lever. It attaches to a peg on the surface of the transfer guide, threads over one roller and under the other, and hooks onto a thin metal peg above the paper path. This peg is directly across from where the green corona wire cleaner for cartridges is stored. Replacement separation belts are typically stored between the corona wire cleaning tool and the edge of the machine. If your machine starts jamming a lot, the separation belt may be broken. You may purchase new belts from your cartridge supplier.

FIGURE 3
The paper separation belt

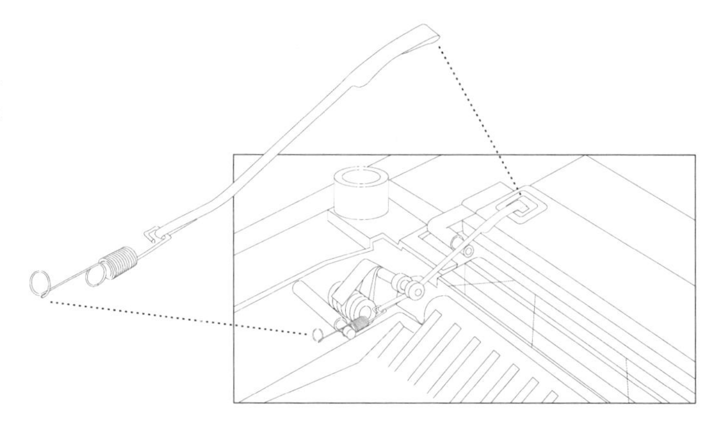

You can tell if the separation belt is dirty just by examining the belt itself. A black smudge on the right side of the page means the separation belt may need cleaning. To clean it, detach the end loop from the metal peg by pulling it gently to the left about two millimeters. Hold that end with one hand, and lift the top roller off the belt. With your other hand, grasp the top of the belt between thumb and forefinger. Gently draw the belt through this delicate "pinch-hold" a few times to clean off any toner smudges. Still holding the end loop, thread it back through the rollers, and hook it on the peg. This requires a little dexterity, and may take some practice before you feel completely comfortable with it. If you use a dampened cloth or cotton swab to clean the separation belt, be sure only to use water. Alcohol or solvents may damage the plastic.

Cleaning the Corona Wires

Vertical black lines on your printed copy indicate that you need to clean the toner cartridge's corona wires as shown in Figure 4, below. These lines may vary from occasional pencil-thin marks to uneven "drops of paint" smeared down the page. Characteristically, they do not appear in the same place on every page.

Remove the cartridge from the printer. Hold it flat, as shown in Figure 4, to avoid spilling toner. Use the green cartridge corona wire cleaning tool which is stored inside the printer, near the latch.

Insert the corona wire cleaner finger into the long slot on top of the cartridge, with the tooth away from the handle. The

FIGURE 4
Cleaning the cartridge corona wire

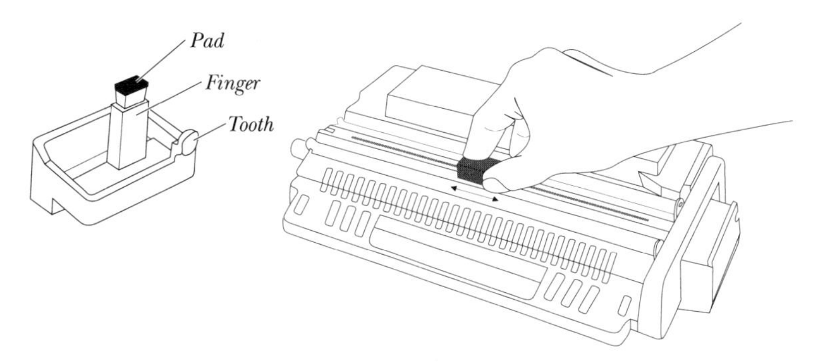

cleaner fits in easily this way and does not work at all if reversed, so you'll know right away if you've placed it correctly. Pop the tool into place, moving it all the way in, until the plastic tool meets the body of the cartridge. You will not damage anything. The flexible plastic film that moves to the side is to protect the drum from light. Slide the tool lengthwise six times, then gently remove it. It may "poing" on removal, but this should not hurt the wire. If the corona cleaning tool looks dirty, it can be cleaned gently with a vacuum.

Wired Wires

Toner streaking can be frustrating. Usually the problem is simply a dirty cartridge corona wire. The corona wire is below the black film that covers the slot in the cartridge. Don't increase your frustration by cleaning the film, which is there simply to protect the cartridge from light. Be sure to insert the corona wire cleaning tool past the film, so that it fits snugly against the cartridge and is cleaning the wire hidden below.

Be sure not to clean any other part of your laser printer, or anything else, with the corona wire cleaning pad, as this will get it dirty so that when you go to clean the cartridge corona wire, you just make matters worse. Vacuum the pad to clean it.

Cleaning the Paper Path

Stains on the back of a page indicate that the paper path may need cleaning. Wipe it with a tissue or soft, barely damp cloth. Be careful of the fine teeth near the corona wires, which can catch your cleaning cloth. These stains could also be a symptom of dirty fuser rollers. The fuser rollers are located under the fuzzy green cover near the paper output tray. Clean them with a barely damp cloth, "pinching" off whatever dust you have wiped up at the end of the roller. Dry them with a clean cloth. Cleaning may also be done with a clean dry cloth, if the roller is not extremely dirty.

Pick Me Up

The paper feed cams, which pick up the paper in laser printers and copiers, are covered with textured rubber that lifts the paper and passes it through the machine. With age and use, the rubber on the cams may harden and lose its grip. Thus, the printer will miss the page pickup, causing the top of the print to appear halfway down the page. Wiping these rollers with acetone may revitalize the rubber and revive the grip. It is important not to use rubbing alcohol, because it will dry out rubber.

Be careful of the fuser roller area—it is hot. The green felt cover diffuses the heat fairly well, but the rollers are uncomfortably hot and could burn you.

Dirty marks on the edges of the paper are almost always caused by a dirty paper tray area. The last area to clean is the paper path under the machine. Start by removing the paper tray and wiping it out. Next, wrap a bit of cotton cloth around your fingers, or using a cloth glove, reach into the paper tray slot under the machine, and wipe all the surfaces down. Move to the other side, open the back door to the paper path, and clean the ridged surfaces of the paper path. Reach in further and wipe the paper feed cams.

By giving your laser printer a regular basic cleaning, you will maintain the high quality of the output that your machine was designed to produce. A little preventive maintenance can go a long way and save you from future trouble.

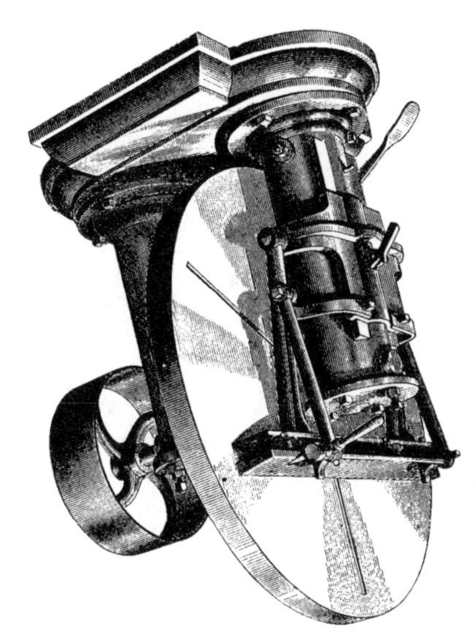

SX Maintenance

Care and Feeding of Your Series II Laser Printer

by Holly Blumenthal of BlackLightning

From *The Flash* Volume 1, Issue 2

The Series II laser printer allows its users to effortlessly print out sharp, beautiful graphics and text. It's easy to become accustomed to this convenience, but when the printer develops ideas of its own, and draws lines or splotches where you never intended them to be, it can be more than a minor inconvenience. Some basic maintenance and knowledge of potential trouble spots can help ensure consistent, clean copy, and increase the lifespan of the machine.

The Series II laser printers, including Apple's LaserWriter II family, Hewlett-Packard's LaserJet Series II, QMS PS810, and many others, are low-maintenance machines. But by the nature of their use, paper dust and toner residue will dirty the machine. There are numerous laser printer cleaning kits on the market that sell for $20 or more. Our research and development department has looked into these products, only to find that they are no better than a few inexpensive items that can be found at a local drugstore. These basic supplies include Q-tips, cotton balls or a soft cloth, isopropyl alcohol, and water. Cleaning supplies should be stored near the laser printer and be easy to access. Once you are familiar with the Series II laser printer, it is easy to periodically check and clean it.

Use the diagram in Figure 1 to familiarize yourself with the basic components of the Series II machines. Turn off the laser

FIGURE 1
Inside the SX laser printer

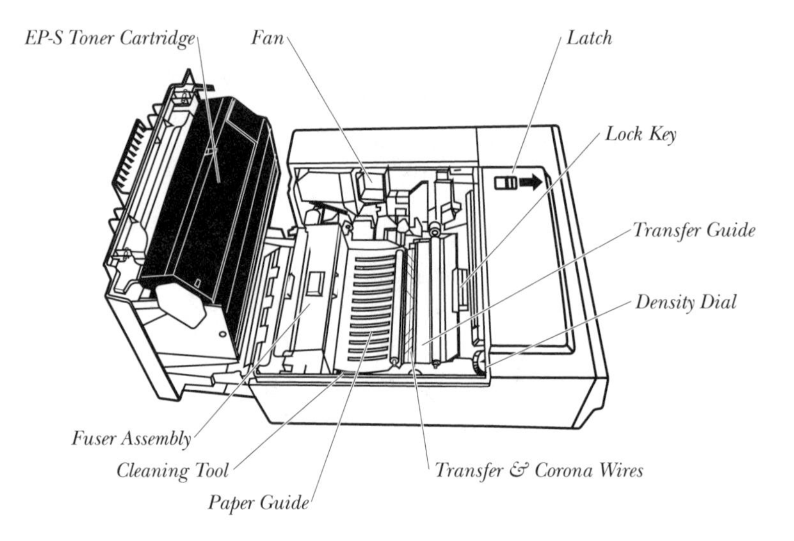

printer before you begin. The button on the top right corner of the printer opens the machine. Push this gently but firmly, allowing the button to slide toward you and then down. Look into the machine to familiarize yourself with the key components of the Series II laser printer: fuser assembly, cleaning tool, paper guide, filter, printer corona wires, discharge pins, transfer guide, lock key, and density dial.

The Fuser Assembly

The fuser assembly is easy to locate. It contains the fuser wand and roller. At the back of the machine, there is a large green cover with the words "Warning: High Temperature". Below this cover is the fuser wand. To remove the wand, lift the cover, grasp the centered fuzzy green handle, and lift. Just below where the wand rests is the fuser roller. Dust particles and toner residue build up on this roller. Gently clean the fuser roller with a cotton ball or a Q-tip, being careful not to scratch it. It is best to try to pick up the dirt, so that it does not fall back into the machine.

We do not recommend using the white square felt on the ends of some fuser wands, as shown in laser printer manuals, because it is more difficult to pick up residue using this method. You are more apt to spill it back into the machine where it will build up, possibly causing problems later. For additional access to the roller, lift the smaller green "back door" located just behind the main fuser assembly lid. In the body of the back door are paper-separation claws which need to be wiped.

A dirty fuser wand may cause vertical streaks down the paper, smearing the print. If the wand has toner caked along the edge, you should get a fresh one. Fuser wands can be purchased from a dealer or remanufacturer. In a pinch, you can clean the felt by scraping it with the edge of a paper clip.

In front of the fuser assembly is the wide black ridged paper guide. Simply clean it with a cotton ball or soft cloth slightly dampened with water. If the back of your paper is dirty, this may be the culprit.

Cleaning the Filter

Just to the right of the paper feed, there is a black hole in the wall of the laser printer. Inside this hole is a filter. Clean the filter gently with a vacuum, soft cloth, or toothbrush. If this filter becomes clogged, the printer will overheat and the filter may need to be replaced. Symptoms of a clogged filter are blurred print, toner bleeding through the paper, abnormally crusty fuser wand buildup, and a burning smell from the machine. It is also important to keep the area around the laser printer clear, to allow adequate airflow to the vents.

The printer transfer wires are just in front of the paper-feed guide. These wires are very fine strands running diagonally across a silver well. There is also a long, fine corona wire running just below the short diagonal transfer wires (Figure 2). Dirty wires may cause white patches, vertical streaks, or light copy. These wires can be carefully cleaned with a Q-tip slightly moistened with isopropyl alcohol. Avoid getting any alcohol on the rollers or plastic parts. Be gentle; these wires are fragile and costly to replace. At each end of the well is an orange pad that can also be wiped off with a Q-tip.

Between the corona wire and paper feed, there is a row of discharge pins. You can use the brush part of the cleaning tool to wipe off the pins. Again, be careful of the corona wires as you do this.

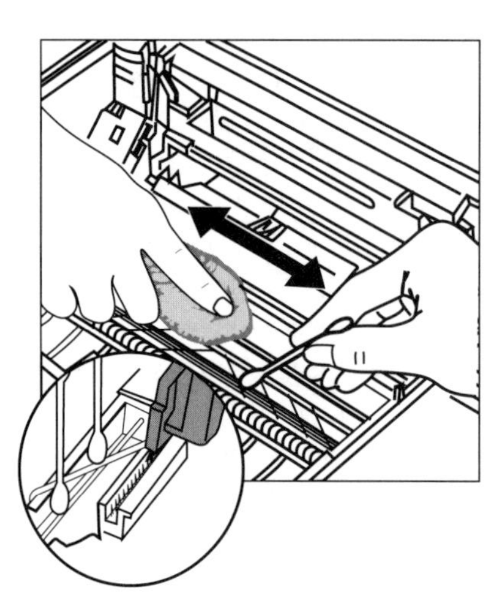

FIGURE 2
Cleaning the transfer wires

The Transfer Guide

Next to the transfer wires is a chrome-colored bar called the transfer guide. Wipe this bar with a slightly damp soft cloth or cotton ball. In front of the transfer guide is a centered green handle. This is the handle of the lock key, which you can lift up. Wipe off any paper dust on the plate below. While cleaning your transfer guide lock key, note the dial to the left. This is the density dial, used to adjust the darkness of your print.

Close the lid, and wipe off the paper tray with a soft cloth or cotton ball. This will help keep your paper crisp and clean.

Care of the cartridge is also very important. Store spare cartridges in a dry, dust-free, cool, dark area. Excessive exposure to light will ruin the photosensitive drum. Be aware that using harsh chemicals containing ammonia, benzene, or other strong agents can cause damage to the cartridge drum. Cleaning the room or office that houses the laser printer should be done with cleaning agents that are less harmful. When opening a new or remanufactured cartridge, be sure to remove the dam before the cartridge is inserted into the machine. This will prevent the possibility of toner falling into your machine as the dam is removed, which can cause black streaks across the page.

Take out the small green cleaning brush (Figure 3) that is just to the right of the paper-feed guide. The padded end of this tool is used to clean the cartridge corona wires. Do not use this pad to clean the machine corona wires, as this will cause the tool to become excessively dirty and prevent it from working properly.

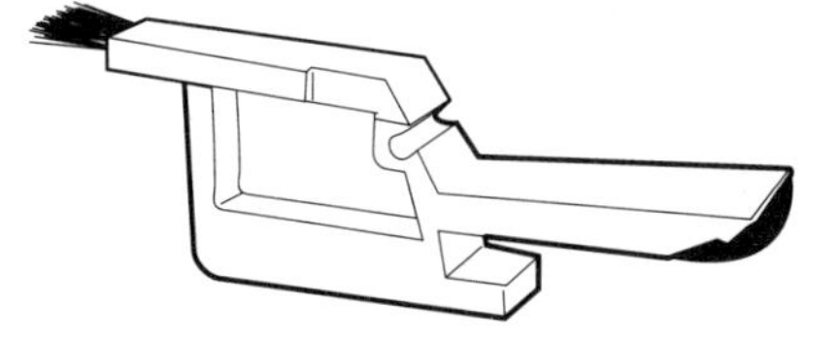

FIGURE 3
Corona wire cleaning tool

Corona Wires

To clean the cartridge corona wires, insert the padded end of the tool into the slot that is covered with black cellophane (note the position of the cleaning tool in Figure 4). Slide the cleaner back and forth in the slot approximately six times, or until the squeaking stops.

What's Black and White...

How to tell if it's the corona wires in the machine or in the cartridge that are causing unsightly streaks? If your laser printer's cartridge corona wires are dirty, black streaks will appear down the printed page. If white streaks or spots appear, the problem lies in the printer. In a copier, white streaks could be caused by either the cartridge or the machine corona wire being dirty. But alas, there is an exception to every rule: EP-L cartridges (H-P IIP) don't have corona wires.

If your copy is light along one side of the page, you may find that rocking the cartridge will help redistribute the toner. This can be helpful if the cartridge has been shipped to you and the

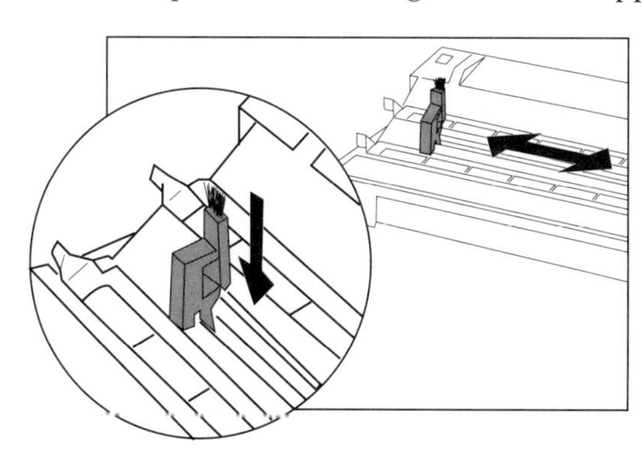

FIGURE 4
Cleaning corona wires

toner has settled to one side in the process. Hold the cartridge flat and rock to 45 degrees, front to back and side to side. If this does not help, a sharp side-to-side shake can be attempted as a last resort. Be sure to clean the corona wires in the cartridge after doing this.

Keeping your laser printer clean and printing properly takes only a small amount of care and a few simple materials. In general, a gentle wipe with a Q-tip or a soft cloth is all that is required. Once you know the areas to key into, a quick glance becomes second nature. A little attention can result in consistently beautiful copy, and add years to the life of the laser printer.
nique may be applicable to making printed circuit board etchant masks.

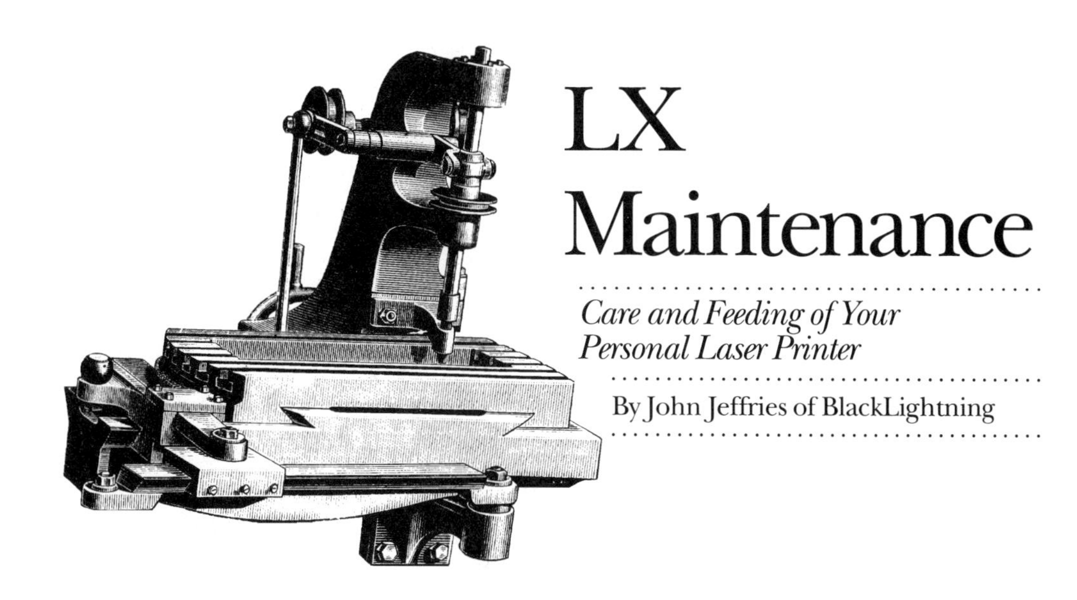

LX Maintenance

Care and Feeding of Your Personal Laser Printer

By John Jeffries of BlackLightning

The Canon LX engine (Figure 1) is used in the latest generation of low-cost "personal" laser printers from many different manufacturers. The printers that use this engine are rated to print about four pages per minute, and usually have a "p" or the word "personal" in their name; some examples are the H-P LaserJet IIp and IIIp, the Apple Personal LaserWriter NT and NTR, and the Canon LPB-4. These printers are called "personal" because they are generally not designed for the rigors of large network printing, but they are priced just right to be used in a small office with a few users.

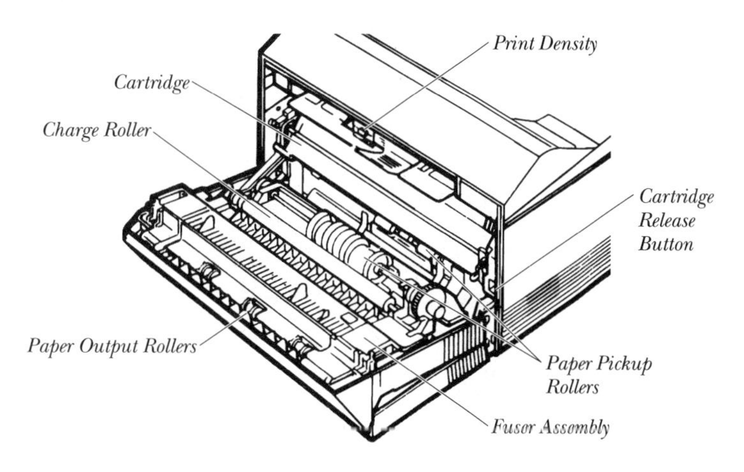

FIGURE 1
The LX engine

The LX engine is the third generation of the Canon laser printer engines following the CX and the SX engines. Canon of Japan actually makes the internal parts of all the printers that use these engines. The printer manufacturers make the cases and provide the interfaces to connect them to your computer. The LX engine continues Canon's trend of making their laser printers easier to use and maintain.

Cleaning Materials

The cleaning materials that you need to clean your personal laser printer are easy to find at most drug stores: Q-tips, a lint-free cloth, isopropyl alcohol, a small vacuum, and acetone. If you keep these supplies near the printer, they will always be handy when you need them. They may even remind you to do the cleaning periodically. It is recommended that the printer be cleaned every time you have to change an empty toner cartridge, or at least four times a year.

Be sure that the area around the laser printer is clean, and clear of obstructions to the airflow. Try to keep the area free of dust, and keep at least two inches clear on all sides to guarantee sufficient cooling. If the printer's cooling vents become blocked, it may overheat and produce poor copy or paper jams. The printer should not be placed in a hot or cold area, or in direct sunlight.

Pulling the Plug

On the LX (EP-L) laser printers, such as the Apple Personal LaserWriters, if there is an erroneous "needs service" message blinking, it may be necessary to turn the machine off and pull the plug before the message will go away.

Whenever you start to clean the printer, be sure that the power is turned off; there are some very high voltages and hot components in the printer. It is best to leave the printer off for half an hour to let the fuser roller cool, so you won't burn yourself. We will start cleaning from the outside, and work into the printer.

Wipe off any dust or dirt that has collected on the top of the printer or in the vents along the sides. If you have a stubborn stain that will not come out, try a mild alcohol or water-based cleaner. It is very important that you do not use anything with harsh chemicals (like ammonia) on the printer, because the fumes can damage the internal parts of the printer and render the drum in the cartridge unusable. If you need to try a new cleaner on the printer, test it on a small, unobtrusive area.

Next, pull down the paper tray on the front of the printer. If the tray is open and there is paper in it, remove the paper so it does not get in the way and does not get dirty. Use your cloth to wipe off anything that has collected on the surface of the paper tray. Be sure to slide out the extension and wipe it, too.

Inside the LX Engine

To open the printer, pull up on the gray release button to the right of the paper opening on the front of the printer. Pull out the cartridge and set it aside, out of direct light. You may have to press on the cartridge release button to get it out. The cartridge requires no cleaning or maintenance for itself, but you should wipe off any accumulated toner or paper dust, so it does not fall into the printer.

Now is a good time to acquaint yourself with the internal parts of the printer. Along the base of the door and in the base of the printer, there are some rubber rollers that pull the paper into the printer. Above these paper pickup rollers in the door is the charge roller. This is the roller that pulls the toner off the photoconductive drum onto the paper. Above the charge roller is the fuser assembly. Be careful; this area may be hot, if the printer was turned on recently. Above the fuser assembly are the paper output rollers. There are two sets of these rollers, one on the inside of the door and one on the outside. As you look into the printer with the door open, you can see the cartridge cavity. You can use your cloth or small vacuum to pick up any dust that has collected there.

Where, Oh, Where...

is that EP-L wand? Nowhere! LX laser printers (EP-L) such as the H-P IIP do not use a fuser wand. So when in the instructions for return shipping admonish "Do not forget to enclose the fuser wand," EP-L users can forget it.

Use a vacuum or cloth to pick up the dust and toner that have collected in all the nooks and crannies in the printer. Be careful not to suck up any loose parts of the printer (there should not be any), and do not knock any delicate areas with the nozzle of the vacuum.

Rejuvenating the Rollers

As the rubber rollers in your printer age, they will begin to harden, to the point where they will no longer pick up the paper. This hardening will cause the printer to jam more often, and to misregister the image on the page. You can delay or even eliminate this aging by occasionally treating all the rubber rollers with acetone. The rollers you should treat are the four paper-pickup rollers at the bottom of the door, and the six output rollers at the top of the door (three on the inside and three on the outside). Do not clean the charge roller with acetone.

This treatment does not need to be done every time you clean the printer, but try to do it at least four times a year. To treat the rollers, dip a Q-tip in the acetone. The Q-tip should be damp, but not dripping with fluid. Next, carefully dab at the rollers with the Q-tip. If the rollers turn easily, then go ahead and spin them so you can get the other side. If the Q-tip comes away black with dirt, then start again with a fresh Q-tip, until the rollers are clean. When you use acetone, be sure to leave the printer open—with no cartridge in it—to let the fumes air out. High concentrations of the cleaner can damage the photoconductive drum in the EP-L cartridge.

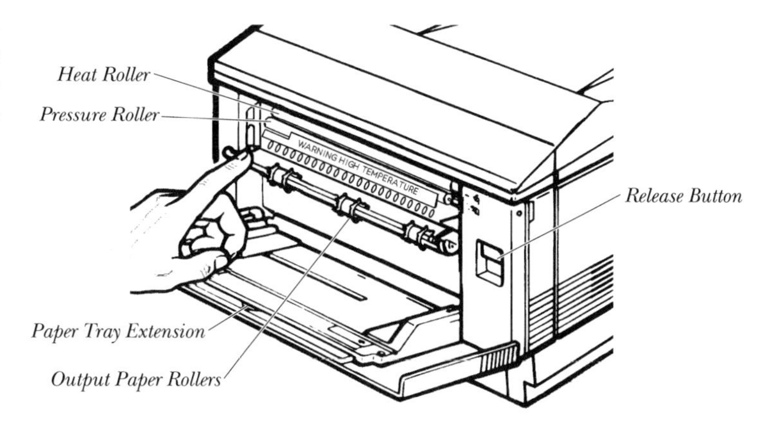

FIGURE 2
The LX fuser assembly

Cleaning the Fuser Assembly

The last area to clean is the fuser assembly. Close the printer, but leave the paper tray down. You can pull down the external fuser access door (Figure 2) to see the fuser rollers. When you open this door, you will see two red rollers running from left to right. The roller closer to you is the pressure roller, made of a soft rubbery material. The roller farther from you is the heat roller. It is much harder, and is the one that gets hot to melt the toner onto the page. When you pull down the access door, you can check the condition of the fuser roller.

If your printouts begin to develop a vertical line on the page, there may be a scratch on the heat roller. If this happens, you will have to replace the fuser roller. We have noticed that a small scratch on this roller does not affect the print quality, but a larger one may be a problem. Because the space is so tight in the fuser assembly, the only way you can clean it is to run the clean page paper through the printer. It is a good idea to do this each time you put a new cartridge into the machine.

That is all that is involved in cleaning your printer. The only thing left to do is to put the cartridge back in the printer and start it back up again. After a few cleanings you will become proficient, and it will not take you more than a couple of minutes to keep your printer in tiptop shape.

Repairing CX Fuser Rollers

Correcting a Common Problem

By John Jeffries of BlackLightning

From *The Flash* Volume 4, Issue 4

Welcome to the first in a series of articles about repairing common laser printers and photocopiers. These articles will take you beyond simply cleaning your machine (see the articles about cleaning the CX, SX, LX, PC for this information) to actually replacing worn or broken parts, and returning an older machine to good running order. With the information in these articles and a few new parts, you should be able to keep your machine in good running order for 500,000 or even 1,000,000 copies. Here at BlackLightning we have several machines that have printed over 500,000 pages, and one that has printed over 800,000 copies. They are still going strong, and we expect to continue using them for years to come.

This article will cover the replacement of the fuser roller, the heater lamp, and the fuser assembly fuse in Series I (CX) laser printers such as the Apple LaserWriter, the LaserWriter Plus and the HP LaserJet. These are the machines that use the EP toner cartridge. This technique can be adapted to the Canon PC desktop copiers.

The fuser roller is one of the first items that will need to be replaced in these printers. It is located at the end of the paper path under a (usually green) cover with warnings about high temperatures. Do not take these warnings lightly; when it is running, the fuser roller heats up to as much as 180°C (356°F), more than hot enough to burn you in short order. Before beginning to work on your machine, leave it turned off for at least an hour to let the fuser roller cool off. It would be best to do this work before you turn the machine on for the day.

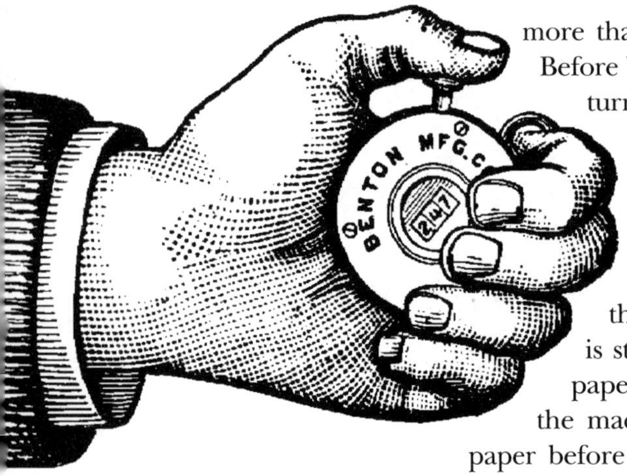

In the machine, the toner is deposited on the paper by the photosensitive drum in the toner cartridge. At this point the toner is still a loose powder, and is only held on the paper by a weak static electric charge. If you open the machine while it is printing and pull out the paper before it gets to the fuser assembly, you will see that the toner just rubs off and makes a mess of everything. After receiving the toner, the paper moves to the fuser assembly, where a combination of heat and high pressure melts and fuses the toner to the surface of the paper.

On exiting the fuser assembly, the leading edge of the paper will tend to stay with the roller and curl up, instead of moving out to the paper tray. To make sure that this does not happen, there are four paper-separation claws that just barely touch the fuser roller; they peel the leading edge of the paper off the fuser roller, and send it to the output tray. It is these claws that cause most of the damage to the fuser roller. If you open the fuser assembly door and the top roller does not look smooth and evenly colored, it probably will need to be replaced in the near future. The scratches will eventually start to leave a ragged line down the page in the same position as the line on the roller. These scratches will be visible on the roller quite a while before they affect the quality of the print. When you notice the scratches, get yourself a new roller, but don't bother doing the replacement until they are causing a problem.

The heater lamp is another part of the fuser assembly that often fails. The heater lamp is a halogen bulb inside the fuser roller that runs the length of the fuser roller. This lamp heats the fuser roller, and will eventually burn out. When you purchase a new fuser roller, also get a new heater lamp. It is a good idea to replace them both at the same time because then you only have to take the fuser assembly apart once. Before it fails completely, a bad lamp can overheat the fuser roller and cause transfer toner to streak.

When you replace the heater lamp in your machine, it is important to get one of the appropriate power. If the lamp is not hot enough, the toner will not melt completely, and it will tend to flake off the page. If the lamp is too hot, it will overheat the roller and cause streaking on the page. This streaking will be especially noticeable if you are using BlackLightning's Transfer Toner. The correct rating for the CX fuser heater lamp is 550 watts.

The final part that may need to be replaced is the fuser lamp fuse. This is a 47 , ¼-watt resistor next to the power supply. This is the least likely to need replacing. If you need another one, you can purchase an identical resistor at your local Radio Shack or other electronics supply store.

Replacement of the fuser roller, heater lamp, and fuse resistor is relatively straightforward and requires a minimum of tools. If you are not mechanically inclined, or are unwilling to accept the risks associated with working on your own equipment, you should have a qualified technician do the replacement. We cannot accept liability for any damage, injury, or loss you may incur resulting from this replacement. Work gently, slowly, and methodically. Be careful not to damage any of the parts, as some of them are fragile and expensive to replace. Please read all directions before starting. When reading these instructions, remember that the near end of the fuser assembly refers to the side where the fuser wand is inserted; the far end is the end closest to the printer's main power switch and the power supply.

While doing this replacement, you will have to deal with several different sizes of screws. It is important that you do not get these mixed up. I like to line up the screws and parts in the order that I removed them, to keep them straight.

Tools Required

- Medium-sized, long, magnetic Phillips screwdriver
- Medium-sized, flat-edge screwdriver
- Soft clean cloth
- Needle-nose pliers

The Procedure

Important! Be sure to turn off and unplug your machine before attempting any work on it.

1. Remove the paper output tray. It is held in place with a tab at each corner.

2. Remove the two screws from the base of the gray plastic end cover on the near side of the fuser assembly. Lift out and place the end cover aside. If it is dirty, wipe the cover gently with a clean, slightly damp cloth.

3. Remove the four screws (two at each end) holding the fuser assembly to the printer bottom (Figure 1). Note that one of the screws may be shorter than the others. In our machine (a LaserWriter Plus), the short screw goes on the near side, farthest from the paper output tray.

FIGURE 1
Fuser assembly attachment screws

4. Lift the fuser assembly part way out. Unplug the two wires, one at either end (Figure 2). Both of these wires are on plugs for easy removal. The needle nose pliers make it easy to get a hold of the wire on the far side; be sure to grab the connector and not the wire. Do not remove the end of the third wire leading from the top of the fuser assembly to the power supply.

FIGURE 2
Fuser assembly wires

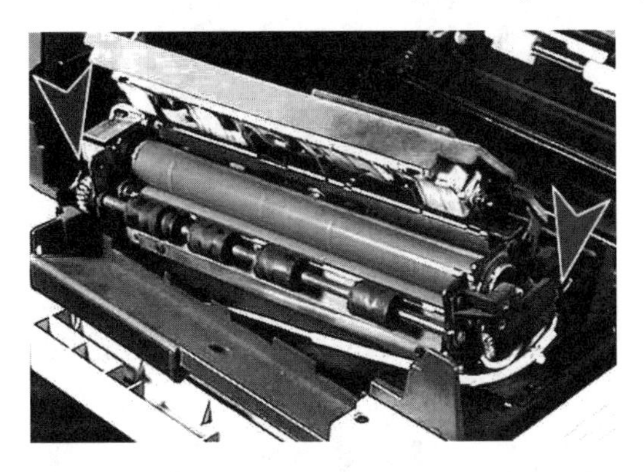

5. Unhook the last wire (leading from the power supply to the top of the assembly) from the silver frame on the far end of the assembly (Figure 3).

FIGURE 3
Power supply wire

6. Lift the fuser assembly out farther. While carefully holding the black lamp holder in place, remove the two screws from the holder on the near side of the fuser assembly (Figure 4).

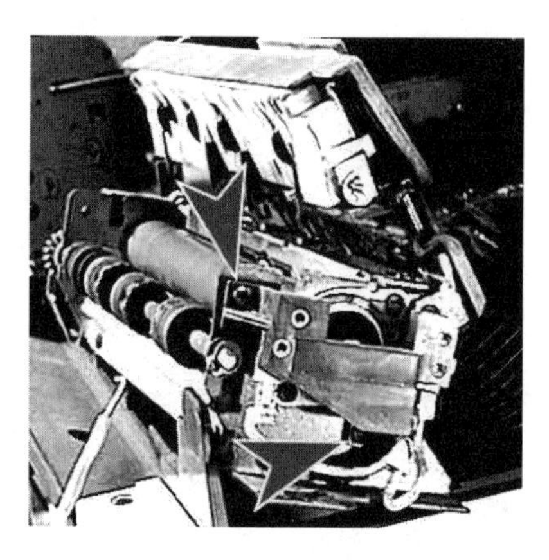

FIGURE 4
Lamp holder

7. While holding the lamp in place, carefully remove the black lamp holder. It works best to reach in from the top with a finger, and push the ceramic lamp end in while removing the lamp holder with the other hand (Figure 5).

FIGURE 5
Removing the lamp holder

8. Gently slide the lamp from the roller and put it aside. Caution! The lamp is fragile. Do not break it. Do not touch the glass part; handle the lamp only by its ends. Save this lamp by wrapping it in a plastic bag. If it has not blown, it may still have some useful life left; it can be saved as an emergency backup in case your new lamp ever blows.

9. Remove the two remaining screws from the silver lamp holder on the far end of the fuser roller (Figure 6). Note the position and attachment points of the attached spring. Do not lose it.

FIGURE 6
The silver lamp holder

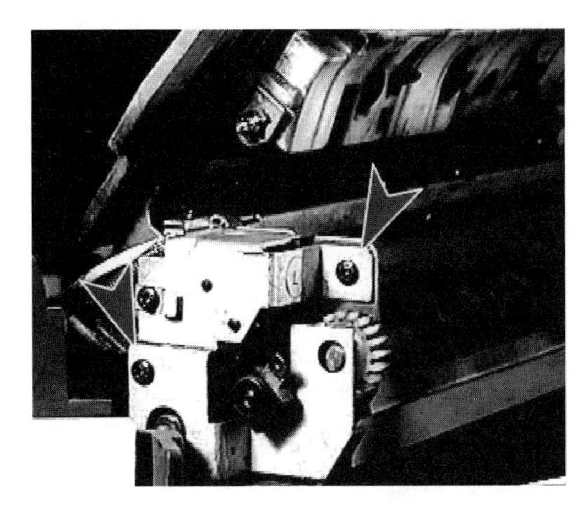

10. Remove the silver lamp holder.

11. Use the flat-edge screwdriver to gently remove the first C-clip from the far end of the fuser roller. If you spread the C-clip too far, it will break; that would not be fun. Be careful not to let it spring off into never-never land. Try to spread it enough to get another tool under the middle; an extra hand helps here.

12. Gently remove the gray metal gear from the far end of the fuser roller.

13. Remove the retaining screws from the rounded, silver, diamond-shaped fuser roller bearing on the near end of the fuser roller. This bearing has a black ring on the inside.

14. Gently slide the roller out of the fuser assembly.

15. Remove the outer C-clip from the near end of the fuser roller.

16. Remove the bearing from the fuser roller. Note the position of the bearing.

17. Remove the third inner C-clip from the near end of the fuser roller. Carefully note the near end of the fuser roller, and identify it on the new fuser roller.

18. Clean the sensor elements in the fuser assembly with a Q-tip and isopropyl alcohol or acetone, to remove buildup.

You have now removed the old, worn parts. It is now time to reverse the process, and install the new fuser roller and heater lamp. Be very careful handling the new heater lamp. Please read the warning that comes with it. The heater lamp is very sensitive to scratches, cracks, fingerprints, and dirt. It should be handled as little as possible, and with great care.

19. Place the third C-clip on the inner groove of the near end of the fuser roller. This end is not slotted, and has two grooves for the C-clips.

20. Place the bearing on the near end of the fuser roller. Remember that the thick part goes on the inside.

21. Place the second C-clip on the outer groove of the near end of the fuser roller.

22. Gently slide the fuser roller back into place in the fuser assembly. Be careful not to scratch the roller. Be sure to get the bearing all the way into the hole in the housing.

23. Replace the two retaining screws on the bearing at the near end of the fuser roller.

24. Replace the gray metal gear on the far end of the fuser roller.

25. Replace the first C-clip on the far end of the fuser roller.

26. Hook the spring back onto the silver lamp holder, and secure the holder with the two screws. That spring is very important, so be sure to get it in the right position.

27. Very carefully insert the fuser lamp from the near end of the fuser roller. Caution! The lamp is fragile. Do not break it or touch the glass part. The nipple half of the lamp should be inserted first. The far end should seat in the copper contact in the silver lamp holder at the far end of the fuser roller. Note that the nipple should point towards the sensors that you cleaned off in step 18.

28. While holding the lamp in place so that it does not touch the inner sides of the fuser roller, replace the black lamp holder, seating the near end of the lamp on

the copper contact of the black lamp holder. Replace the two retaining screws for the lamp holder.

29. Reattach the wires to the fuser assembly.

30. Reseat the fuser assembly in the printer.

31. Replace the four screws holding the fuser assembly to the bottom of the laser printer. The short screw goes in the near side, closer to the transfer wires. If during testing you find that you get paper jams as the paper enters the fuser assembly, then the position of the assembly must be adjusted. Do this by loosening these four screws and moving it.

32. Replace the end cover and the two retaining screws on the near end of the fuser assembly.

Testing

Place a fuser wand and cartridge in your laser printer, close it, plug it in, and turn it on. The green light should flash. If the green light fails to go on through the warmup period or does not come on at all, turn off the laser printer and check all the connections to the fuser assembly. Retest, and if this does not solve the problem, the fuse in the power supply beside the far end of the fuser assembly may need to be replaced.

Replacement of the Heater Lamp Fuse

This is almost never necessary, but the fuse can be purchased at Radio Shack or any good electronics store. The fuse is a socketed 47 , ¼-watt resistor. It is important to get one of this value; if it has a higher power rating, it will not burn out when it is supposed to, and this could cause damage to other parts of the machine. This should only be attempted if you are familiar with soldering and the ohmmeter.

Tools

- An ohmmeter or continuity meter
- A small soldering iron
- Solder
- Phillips screwdriver

Procedure

1. Turn off and unplug your laser printer.
2. Remove the screw from the far side of the power supply (Figure 7).

FIGURE 7
Power supply cover retaining screw

3. Remove the power supply cover.
4. Remove the fuse plug from its socket on the underside of the upper circuit board (Figure 8).

FIGURE 8
Removing the fuse plug

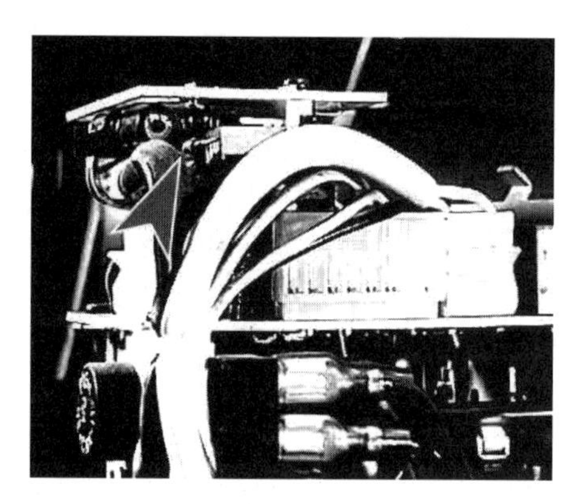

5. Check the resistance of the fuse. It should be 47W. If the circuit is open, replace the fuse by clipping out the old resistor and soldering in the new resistor.

6. Replace the fuse and holder in the socket on the circuit board.

7. Replace the power supply cover and its screw.

8. Test the laser printer again.

This concludes the replaccment of the CX fuser assembly parts. With a bit of mechanical finesse, caution, and careful work, your printer is ready for tens of thousands more pages of quality printing.

Repairing SX Fuser Rollers

Fix Your Printer Yourself and Save Money

By John Jeffries of BlackLightning

Welcome to Beyond The Manual, where we feature articles about maintaining and repairing lasers printers. We will take you from cleaning your machine all the way to replacing worn or broken parts and returning an older machine to good running order. With the information in these articles, some cleaning supplies and a few new parts, your machine may give you a million or more copies. Here at the offices of *Flash* magazine, we have several machines that have printed over 500,000 pages, and one that has printed over 900,000 copies!

This issue covers the replacement of the fuser roller and the heater lamp in printers based on the Canon SX engine. This engine was used in the H-P LaserJet II, H-P LaserJet III, most of the Apple LaserWriter II's, and many other popular printers using the EP-S toner cartridge.

When the fuser roller in your printer needs to be replaced, you have several options. Each of these options represents a trade-off between cost, time, convenience and warrantee.

- You can buy the new parts from a mail-order company for less than $100 and do the repair yourself in about an hour.

- There are mail-order companies that will replace just the fuser roller for between $150 and $200, including the shipping.
- You can remove the fuser assembly, as outlined below, and trade it for a new or rebuilt assembly available mail-order for about $225.
- You can take the printer to your local dealer and, for $250 to $550 in parts plus a half hour's labor, they will do the repair for you.
- Finally, you could discard the the old printer and, for upwards of $1,000, buy a new printer.

The fuser roller is one the of the first parts that will need to be replaced in these printers. Canon claims that it will last about 100,000 pages. To find the fuser assembly, open the printer as though you were going to change the cartridge. Lookiong toward the back of the printer, find the fuzzy green lid that says "Warning! High Temperature." When you lift the fuzzy green lid to replace the fuser wand that comes with each cartridge, you can just see the fuser roller. You should heed the high temperature warnings; if the printer is running, the fuser roller is heated to about 180°C (356°F). Before beginning work on your printer, leave it turned off and unplugged for at least an hour to allow the fuser roller to cool.

In the printer, the toner is deposited on the paper by the photosensitive drum in the toner cartridge. At this point the toner is still a loose powder and is held in place with just a weak static electric charge. If you open the machine while it is printing and pull out the paper before it gets to the fuser assembly, you will see that the toner just rubs off and makes a mess. After receiving the toner, the paper moves to the fuser assembly where a combination of heat and high pressure fuses the toner to the surface of the paper. On exiting the fuser assembly, the leading edge of the paper will tend to stay with the roller and curl up instead of moving out to the paper tray. To make sure that this does not happen, there are four paper separation claws that just barely touch the fuser roller; they peel the leading edge of the paper off the fuser roller and send it out to the output tray. You can see the claws by flipping down the back door of the fuser assembly. It is these claws that cause most of the damage to the fuser roller.

If you open the fuser assembly door and the top roller does not look smooth and evenly colored, it probably will need to be replaced in the near future. The scratches will eventually leave a ragged line of toner down the page in the same position as the line on the roller. These scratches will be visible on the roller for some time before they affect the quality of the print. When you notice the scratches, get yourself a new roller. You can wait to do the replacement until the scratches show up on the page. Periodically scraping off the crust that accumulates on the tip of each paper separation claw will extend the life of the fuser roller.

The heater lamp is another part of the fuser assembly that often fails, although less often than the fuser roller. The heater lamp is a long halogen bulb inside the fuser roller. This lamp heats the fuser roller to melt the toner, and will eventually burn out. When you purchase a new fuser roller, get a new heater lamp. It is a good idea to replace them both at the same time because then you only have to take the fuser assembly apart once. If the old lamp has been working fine, save it in the box the new lamp came in for an emergency.

Before it fails completely, a bad lamp may overheat the fuser roller and cause streaks on the printed page. When you replace the heater lamp in your machine, it is important to get one of the appropriate power. If the lamp is not hot enough, the toner will not melt completely and it will tend to flake off the page. If the lamp is too hot, it will overheat the roller and cause streaking on the page. The correct rating for the SX fuser heater lamp is 650 watts for the printers with a 115V power supply and 570 watts for the printers with a 240V (European) power supply. When working with the heater lamp, be aware that it is a halogen bulb. If you leave any fingerprints on it, they will cause the glass to overheat and even explode when the lamp is turned on. If the lamp gets dirty, clean it carefully with a clean cloth or a cotton ball and isopropyl alcohol.

Replacement of the fuser roller and heater lamp is relatively straightforward and requires a small number of basic tools. If you are not mechanically inclined or are unwilling to accept the risks associated with working on your own equipment, you should have a qualified technician do the replacement. Neither the author, the editors of *Flash* magazine nor the publisher can accept liability for any damage, injury, or loss you may incur resulting from this replacement. You are doing this at your own risk. Work gently, slowly, and methodically. Be careful not to damage any of the parts, as some of them are fragile and expensive to replace. Please read all directions before starting.

When reading these instructions, remember that right refers to the side closest to the power switch when the assembly is in the machine. Left is farthest from the power switch. Front is the side nearest the paper input tray. The fuser wand is inserted into the top of the assembly. This orientation continues to apply after you have removed the assembly from the printer (Figure 1).

While doing this replacement, you will have to deal with several different sizes of screws. It is important that you do not get these mixed up. I like to line the screws and parts up in the order that I removed them to keep them straight. Another method is to put each screw a few turns into its hole after you have removed a part.

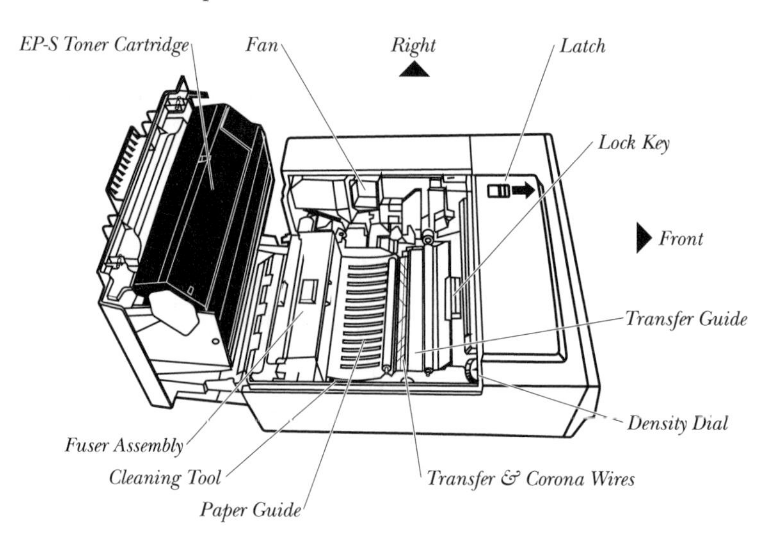

FIGURE 1
SX Printer orientation

Tools Required

- Medium sized, long, magnetic philips screwdriver
- Small flat edged screw driver
- Soft, clean cloth
- Needle nosed pliers

The Procedure

Important! Be sure to turn off and unplug your machine before attempting any work on the machine. Even if you do not electrocute yourself, there is a good chance of frying the printer's circuits if you do not take this precaution.

1. Remove the input paper tray and take the fuser wand out of the fuser assembly so it does not fall out later.

2. Remove the four screws holding the fuser assembly to the chassis of the printer. All four screws are identical. The two front screws are easy to find, but the back ones are just barely visible. The right rear screw is in line with and to the right of the gear on the top of the fuser assembly. The left rear screw is in line with that gear and just in front of the large spring in the printer door (Figure 2).

FIGURE 2
Fuser assembly attachment screws

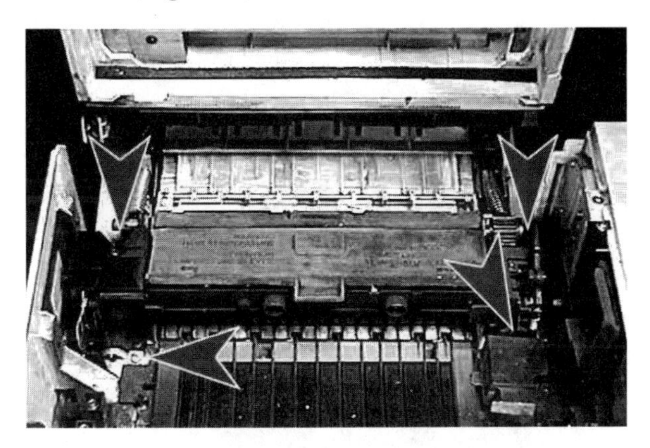

3. Lift the assembly straight out and set it on a table with good light and enough room to lay out the parts—about two feet by two feet. The right side will probably come out easily, but the circuit board connector on the left side may be more difficult. Rock it slightly front to back if it resists. If it is very hard to pull out, you may have removed the wrong screws. Check your work, and try again.

4. Remove the plastic covers from the left and right ends. Each one is held on with a screw on the top of the

assembly. The right cover is a tight fit to squeeze past the lever. If you wiggle it gently, it will come off (Figure 3).

FIGURE 3
Fuser assembly plastic covers

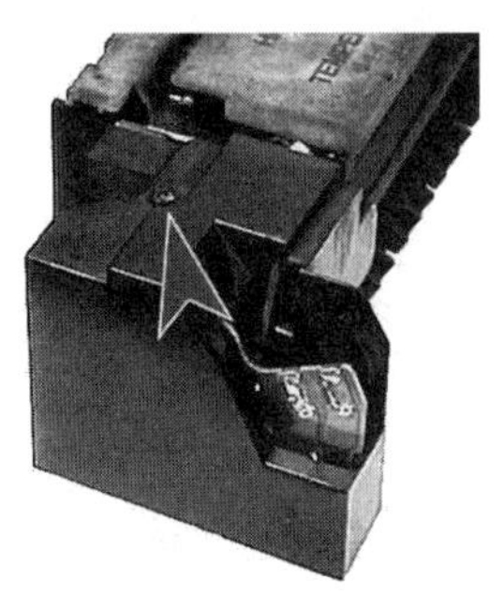

Left plastic cover

Right plastic cover

5. Pry off the C-clip that holds the large gear on the top of the right end, and pull the gear off its post. When removing these C-clips, be very careful not to let them spring off into never-never land. You may not find them again.
6. Remove the spring from the lever and the plastic post on the right side (Figure 4).

FIGURE 4
Spring and thermister contact

7. Lift the fuser cover and note how the wire that leads from the left end to the middle is routed under the thermister cover tabs. Remove the four screws that hold the thermister cover to the front of the fuser assembly, one at each end and at the one third and two thirds points. Pull the cover forward until its tabs clear the top of the assembly, then lower it to get free of the

thermoprotector. These are the components that measure and control the temperature of the fuser roller. Be careful not to damage them or their contacts.

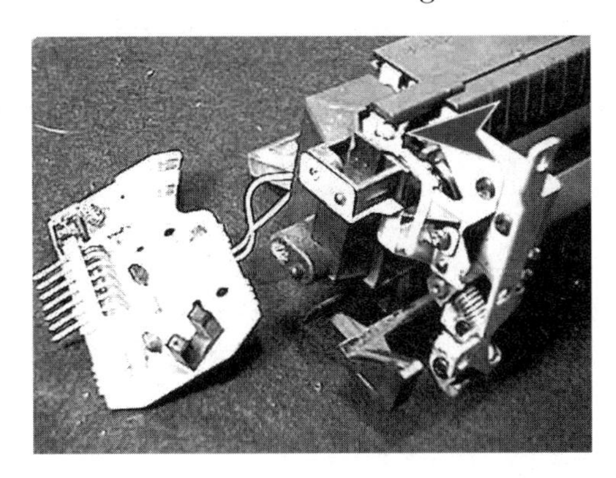

FIGURE 5
Thermister assembly

8. Unscrew the small circuit board from the left end of the assembly. Be careful not to damage the wires that lead to the thermister. The contacts on the circuit board could break with too much flexing (Figure 6).

FIGURE 6
Circuit board on left end

9. Tip the assembly back and remove the plastic paper guide from the front of the assembly below where the thermister cover was. There are two screws holding the paper guide in place.

10. This step is only necessary if you plan to replace the heater lamp; otherwise skip to step 11. Be careful not to break the plastic parts in doing this step. Set the fuser assembly upside down on its fuzzy green cover. Locate the white heater lamp connector on the right end of

the assembly. Find the wire leading from the connector to the heater lamp. Hint: the loose wire is the thermister connection that you removed in step 7. The wire is held in the connector with a plastic tab on the outboard side. Reach the small, flat screwdriver into the outboard side of the open end of the connector. While pushing the wire into the connector with your other hand, push the plastic tab out of the the way of the metal connector on the end of the wire. With the tab out of the way—you have to hold it there or it will spring back—you can pull the wire out of the plastic connector. Now set the assembly back right side up (Figure 7).

FIGURE 7
Thermister connection wire

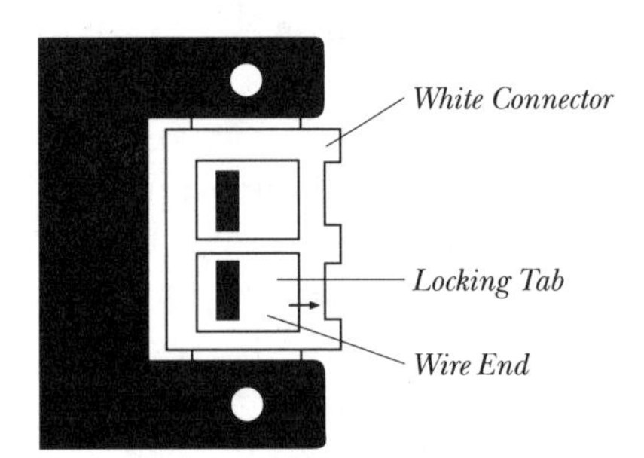

11. Pry off the C-clip that holds the double gear on the lower, right, front corner of the assembly. Push the post back through the gear and remove the gear and the hanger. Be careful not to distort the small spring connected to the top of the hanger. The spring can be disconnected from the hanger and left hanging on the plastic post, but it may be easier to leave the whole assembly together (Figure 8).

FIGURE 8
Lower double gear

12. Remove the plastic end cover that holds the top lever and the heater lamp connector on the right end. Be careful with it, as the plastic is easily broken. If it breaks, you can repair it with super glue.

13. Remove the similar plastic cover on the left end. Don't lose the metal piece that goes from the front circuit board screw to the screw that holds the plastic cover to the frame of the assembly.

14. While holding the heater lamp in place by its ceramic end, remove the right side lamp holder. Carefully pull the lamp out of the fuser roller, disengaging the wire on the left end from the lamp holder on that side as you go. As you pull the lamp out, do not twist it. There is a bulge in the middle of the lamp that must point towards the thermister. Note the orientation of the bulge as it exits the fuser roller. Finally, try not to scrape the lamp against the inside of the roller as you remove it (Figure 9).

FIGURE 9
Right lamp holder

15. Remove the lamp holder, the fuser roller electrical contact, and the contact ring from the left end of the assembly. The lamp holder and the contact are each held on with one screw. The contact ring is just a press fit in the roller (Figure 10).

FIGURE 10
Right lamp holder

16. Remove the C-clip that holds the small gear at two O'clock on the right end. Remove the ring clip that holds the gear on the end of the fuser roller. Remove the identical clip on the left end of the roller. The easiest way to get these clips off is to spread the ends of the clip and insert the flat screwdriver under the opposite side of the clip. From there you can carefully pry it off. Be careful not to spread the clip too much; it may break. Again, be careful not to let the clip spring off and disappear (Figure 11).

FIGURE 11
Right gears

17. Pull the small gear off its post and the big gear off the right end of the roller. Note that the wider shoulder is on the inboard side. Note also the slot in the roller that engages a tooth on the inner surface of the gear. When installing the new roller, be sure to put the slot on the right side.

18. Remove the two screws holding the right bearing in place shown in Figure 11. Pull the right bearing out of

its hole. Note that the shoulder is again wider on the inboard side. It is important to put it back together in the same orientation.

19. Pull the fuser roller out from the right. It will be hard to pull at first because the lower pressure roller pushes up so hard on it. Do not try to release the pressure on the lower roller; you will never get it right again. Just pull the top roller out firmly. You can use a pair of pliers to sqeeze the spring on the left end of the assembly. This will make it easier to slide out.

We have now completed the disassembly of the fuser assembly. The next step is to clean off any parts that are dirty and reassemble it with the new parts. Reassembly is essentially the reverse of disassembly.

20. Clean the accumulated crust off the thermister with a Q-tip or cotton ball. Ascetone, isopropyl alcohol or a citric distilate should work well as a solvent. Be gentle with these parts as they are fragile, but they need to be as clean as possible to extend the life of the new roller (Figure 12).

FIGURE 12
Dirty thermister

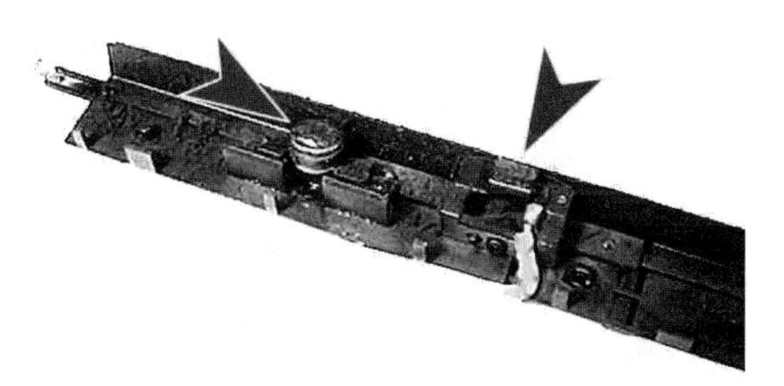

21. Insert the new roller in the right end of the assembly. The notch should be on the right side. Because the lower roller presses so hard on the upper roller, it will be hard to push the roller through. Do not try to force the roller in, as this will surely mar the lower pressure roller and affect the quality of your output. You can use a pair of pliers to sqeeze the spring on the left end of the assembly. This will make it easier to slide in. When

inserting the roller be careful not to scratch its surface on the right side bearing hole.

22. Install the bearing on the right side with the thicker shoulder on the inside. Use two screws to hold the bearing in place.

23. Replace the small gear on its post and the larger gear on the fuser roller. Be sure to align the inner tooth with the slot in the fuser roller.

24. Install the C-clip on the small gear and one ring clip on each end of the fuser roller. Make sure that each clip snaps into its groove; otherwise it will fall off during use.

25. Install the ring contact, the contact brushes, and the lamp holder on the left side.

26. Insert the heater lamp into the fuser roller. The end with the circular connector on the end of its wire should be on the left side of the assembly. Be sure the bulge on the lamp points toward where the thermister will be. Hook the end of the lamp on the left lamp holder.

27. Install the right lamp holder and hook the lamp on it.

28. Install the plastic end cover on the left end. There is a felt pad that cleans the contact ring in the roller. If the roller has been squeaking, a dab of silicone lubricant on this pad may reduce the noise.

29. Install the plastic end cover on the right side with the metal gear hanger.

30. Plug the heater lamp contact into the the white plastic connector on the right if you disconnected it in step 10.

31. Install the thermister cover on the top of the front side of the assembly. The left lamp contact wire should loop down under the end cover and up to the screw that holds the left side of the cover in place. The wire from the white connector on the right should loop up over the end cover and down to the screw that holds the

right side of the cover. Be sure to get the top wire under all the tabs along the top.

32. Put the circuit board on the left end with its two screws. You may have to swing the large plastic paper sensor out of the way to get it to sit properly.

33. Install the lower double gear on the right side the same way it came out. The smaller gear goes on the left. Be sure to get the metal tab of the hanger to the rear of the lever. The lever is supposed to disengange the gears from the main drive when the printer is opened.

34. Install the paper guide below the thermister cover with its two screws.

35. Connect the spring between the lever and the plastic post on the right side.

36. Install the big gear on the post on the top right side of the printer.

36. Replace the plastic cover on each end of the assembly with one screw on each side.

37. Install the assembly into the printer. Check to make sure that the connectors line up properly and are fully seated. Secure the assembly with the four screws.

38. Insert the paper tray, plug in the printer and run a few test pages to make sure everything works. Your task is complete.

This concludes the replacement of the SX fuser assembly parts. With a bit of mechanical finesse, caution and careful work, your printer is ready for many more pages of quality printing.

Ozone

Friend and Foe

By Holly Blumenthal of BlackLightning
From *The Flash* Volume 4, Issue 1

Along with providing beautiful copy and graphics, your laser printer and photocopier also emit ozone. This relatively fragile chemical, which is such an important presence in our atmosphere, is not so welcome closer to home; in large quantities, it can be a health menace.

The ozone layer, which envelops our earth 13 to 35 miles above the surface, is necessary for life on this planet (Figure 1). This relatively thin blanket absorbs much of the sun's harmful ultraviolet light and is the only substance to do so. The ozone layer also helps keep the weather confined to the atmosphere near the earth's surface, preventing temperatures from taking a permanent plummet.

FIGURE 1
Ozone in the atmosphere

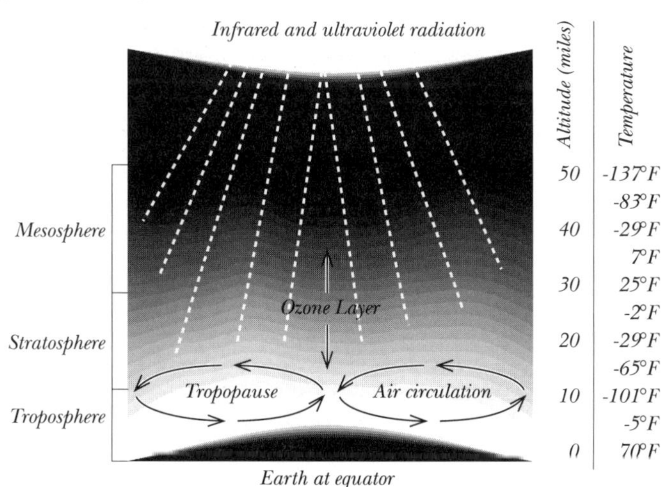

FIGURE 2
Ozone levels

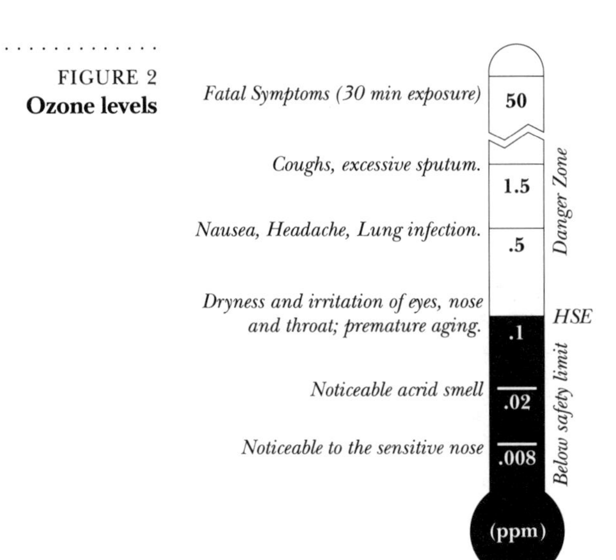

Closer to the earth, where we live, this same chemical can be a health hazard (Figure 2). Ozone contributes to smog, and can cause symptoms ranging from a dry nose and throat to (in sufficient quantity) death. The British Health Safety and Executive (HSE) Guidance Note EH38 (published in 1983) recommends an ozone exposure limit of 0.1 parts per million (ppm), averaged over an 8-hour day. Even at this level, one may experience dryness and irritation of eyes, nose, or throat, and possibly premature aging. Nausea, headache, and increased risk of lung infection may occur at an exposure level of 0.5 ppm. Extremely high exposure of 50 ppm for 30 minutes can be fatal.

How is ozone formed? As shown in Figure 3, oxygen molecules normally have 2 atoms (O_2). Ozone is simply an oxygen molecule with an extra atom (O_3). Sunlight or discharges of static electricity will split oxygen molecules, forming single oxy-

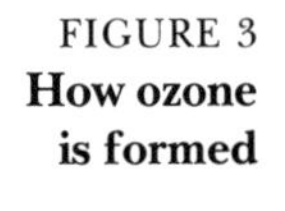
FIGURE 3
How ozone is formed

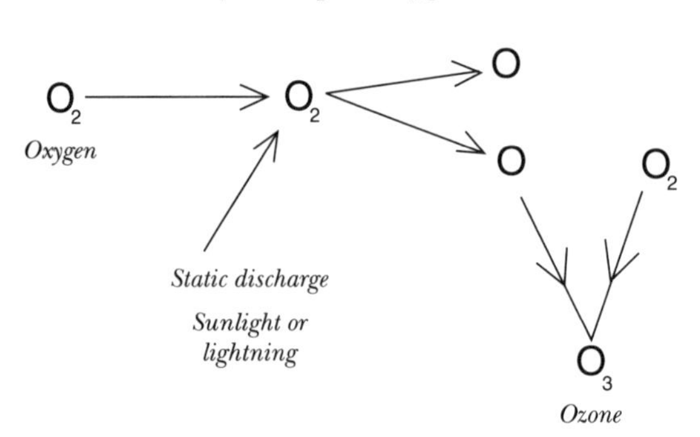

gen atoms (O). These single atoms (O) link with oxygen molecules (O_2) to form ozone (O_3). Ozone molecules are unstable and easily broken down, but during that process, the ozone will attack just about everything except glass and some stainless steel. Ozone is even used as a commercial bleaching agent. So it is not surprising that while important to our atmosphere, direct ozone exposure may be hazardous to your health.

Ozone from Laser Printers

The same electrostatic charges that produce ozone are used by printers and photocopiers to charge the drum. This attracts the toner from the reservoir to the photostatic drum. Then the toner is pulled down onto the paper, using more electrostatic charges.

There are filters built into printers to address the creation of ozone. The activated carbon of these filters breaks the ozone down into oxygen (O_2) molecules before it leaves the printer. New printers emit ozone in levels far below the HSE-recommended limit of 0.1 ppm. However, the ozone filters built into printers become less effective as they become clogged with dust from the air, paper, and toner. After 12 to 18 months, it is prudent to replace the filter or add an external filter. Beware that clogging is faster if room ventilation is poor, or if there is a large amount of dust in the air.

In the Series I laser printers, pull the cartridge out to find the filter. Look into the cartridge cavity in the lid. Toward the top and left of the machine is a rectangular charcoal filter. There is a matching series of horizontal vent slots on the outside of the machine, just above the paper tray. In the Series II machines, the filter is located in the base. There is what looks like a square black cavern in front of the green "Warning: High Temperature" lid, on the inner right wall of the printer. This cave is where the filters and cooling fan are located. Just above the power switch of the Series II printer are the corresponding horizontal vent slots. Vacuum these filter and vent areas thoroughly to remove dust. This allows the filters to work more effectively.

It is important to place printers as far away from work areas as is efficient. Be sure that the vents are not directed toward a work area. Ventilation is important, both to decrease the concentration of ozone in the office and to increase the longevity of the printer's filters. Replace filters annually or every 25,000 to 50,000 pages, whichever comes first.

A little trick we use with our laser printers is to apply a small label to the ozone filter with the date and number of copies at which the filter was changed.

It is difficult to know when a filter is clogged. In this case, it may be your nose that knows. Derived from the Greek word *ozein*, "to smell,"ozone at ground level gives off a pungent, acrid odor. You may have noticed this aroma around high-tension power lines, electric toy trains, or after a lightning storm. The ozone smell is noticeable at levels well below the HSE-recommended limit of 0.1 ppm. A concentration as low as 0.008 ppm can be detected by sensitive noses, and almost anyone can pick up the acrid odor at levels of 0.02 ppm. So being able to smell ozone from your printer may be cause for concern, but not for panic.

Replacement Filters

Replacement filters are available for the Series I (EP engine) laser printers and the PC10–PC25 copiers. An external snap-on filter is available for the Series II (EP-S engine) laser printers. When ordering a filter, be sure to specify:

1. Type of machine: EP, PC, or EP-S
2. Manufacturer: Hewlett-Packard or Apple
3. Age (EP-S only)

- prior to 7/89 (does *not* have green duct door inside printer on right)
- after 7/89 (*does* have green duct door inside printer on right)

A place for everything, and everything in its place. Miles above us, the ozone layer is an important protective blanket, but in our homes and offices, ozone can be a menace to our health.

Further Reading

Buyers Laboratory. "Ozone: Hidden Danger in Copiers?" *Copy Magazine,* September 1991.

Fox, Barry. "Safety Body May Strengthen Ozone Controls for Offices." *New Scientist,* April 7, 1990.

Gribbin, John. *The Hole in the Sky.* Bantam Books, New York, 1988.

Roan, Sharon L. *Ozone Crisis.* John Wiley & Sons, Inc., New York, 1989.

Theroux, Paul. *O-Zone.* Ballantine Books, New York, 1986. Speculative in nature.

The American Challenge

Antitrust Laws, You, the Dealer, and the Law

By T. Fable Of BlackLightning

The next time your printer or copier serviceperson says this:

"Sorry, I'll have to void the warranty on your equipment because you're not using the OEM's brand of toner," or: "I'll have to charge you for this service call, because you're not using the OEM's brand of toner."

Do this:

Inform your service person that it is illegal to require or force the owner of equipment to use the original equipment manufacturer's (OEM) brand of supplies. To make this a requirement is in violation of the Sherman Clayton Antitrust Acts.

The IBM Precedent

A classic example of this issue was brought before the US Supreme Court, involving IBM versus the United States. IBM leased data-processing machines to customers, with the requirement that they only use the tabulating cards manufactured by IBM. Their customers were threatened with termination of their lease if they used cards produced by other manufacturers. As decided by the US Supreme Court, this requirement was held to constitute a "tying agreement," and was found to be in violation of the antitrust laws.

Don't be intimidated by sales or service people. Let them know that a copier manufacturer cannot legally require, in writing or verbally, that a copier owner or lessee purchase supplies exclusively from them. In order to make this kind of requirement, they must conclusively demonstrate and prove that other brands are incompatible with their equipment.

You have the right to use the products of your choice. Inform your service representative that they have no grounds to tell you otherwise!

Show this information to anyone who insists on voiding a warranty or charging you for a service call because you weren't using the OEM's brand of toner. Protect your rights to freedom of choice in choosing your vendors.

OEM's Recycle!?!?

The Big Players Get on the Bandwagon

By Catherine Croft of BlackLightning

Hewlett-Packard, Canon, and many other vendors of the popular EP, EP-S, and EP-L machines are now encouraging their customers to have their toner cartridges remanufactured. Finally, the big boys have admitted what we've all known for years—recycling your empty toner cartridges makes environmental and economic sense. However, this doesn't mean that there is any price reduction from H-P in the works, so remanufacturing with the smaller players still makes the most sense, and will save you dollars while you save the environment.

In fact, not only will you save more money by buying remanufactured toner cartridges from a high-quality remanufacturer, but you will also have a wider selection of cartridges and be doing more for the environment than if you give them to H-P or Canon. Why? Because H-P is melting down and recycling some of the cartridge parts; they are not remanufacturing.

The environmental credo goes like this: "Reduce, Reuse, Recycle!" *Reduce* means lowering the amount of materials that you use—conserving your resources. *Reuse* refers to remanufacturing—making something continue to serve you, rather than buying new. *Recycling* is the last-ditch effort before you add it to the landfill—breaking it down to its components, and making something new. Remanufacturing is better because it wastes less materials, and uses less energy to make the cartridge ready for use again. When you melt the parts down, you lose material and waste energy, compared with simply reusing the parts.

H-P has stated in a letter to *The Flash* that "the use of non-H-P toner does not affect your warranty." Dealers and service technicians are not authorized to void a warranty because remanufactured toner cartridges were used. This validates what BlackLightning has been saying all along: "You have the right to use the products of your choice!" Don't let anyone tell you otherwise. This applies to Hewlett-Packard and all other vendors of supplies and equipment. It's the law.

Remanufacturing

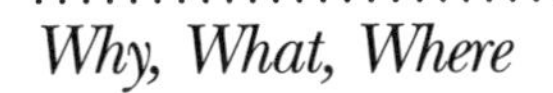

Why, What, Where

By Walter Vose Jeffries of BlackLightning
From *The Flash* Volume 2, Issue 1

By having your cartridges remanufactured, you can save hundreds and possibly thousands of dollars per year, get more copies per cartridge, higher-quality copies, and interesting specialty toners—all provided you can find the right remanufacturer. The smart consumer has only to remember that all remanufactured cartridges are not created equal.

You may buy your cartridges from Canon, Apple, Hewlett-Packard, or QMS, who may have manufactured your printer or copier, but whomever you buy from, the cartridges all come from Canon's factory in Japan. Canon makes the print engine used in most popular laser printers, standardizing the cartridge across value added resellers such as Hewlett-Packard and Apple.

Canon's ingenious system significantly decreases the typical number of service visits per year by placing the toner, light-sensitive drum, developer roller, and other critical components in a "disposable" cartridge. Thus, when you replace your laser printer's cartridge, you are—in effect—performing a service call by replacing these parts. This convenience and reliability comes at a cost: the high price of cartridges. New cartridges, which carry a suggested price of $130.95, may range from $99 to $169 depending on the source, type, availability, and quantity that you purchase.

Despite being marketed as disposable, these cartridges are reusable, when properly remanufactured. At costs far below the cost of a new one, you can reuse your old cartridge dozens of times, even indefinitely if you have critical components replaced periodically as they wear, just as you replace the tires on your car when they become bald.

Typical remanufactured black cartridge prices range from $49 to $99, depending again on source, type, quantity, and most importantly, quality. By finding a reputable remanufacturer, you will realize more than just cost savings. A reputable remanufacturer completely services the cartridge and refills the reservoir with more, higher quality toner than Canon does. Thus, you change cartridges less frequently and get better looking copy.

Additionally, with remanufactured toner cartridges you can get a wide variety of toners that are unavailable from the OEM, be it Canon, Apple, or H-P. These include specialty toners for finer text, darker graphics, high-resolution printers, MICR toners for check printing, colors, and sublimation heat-transfer toners. You can also get Emerald drums, which will make a cartridge last for many years—as the original photoimaging drum in the cartridge is the most likely component to fail.

Breaking In Is Hard to Do

First impressions are important, but there are exceptions to every rule. Never judge the toner quality of your cartridge by the appearance of the first few pages. All toner cartridges need a break-in period in order to produce the best possible copy. The good news is that remanufactured cartridges usually have a break-in period of 50 pages or less, as opposed to new cartridges which sometimes need as many as 200 pages run off to produce their best copy.

When searching for a reputable remanufacturer to do business with, take the time to survey the market. Check out advertisements in the backs of magazines. Talk to other people who use laser printers. Call up a few of the most promising companies; many will have toll-free 800 numbers. Develop your own criteria for rating the companies. The following will give you a start.

How long have they been in business? The longer, the better. There have been a number of fly-by-night and quick-start companies in recharging, who start up with the idea that this is something anyone can do simply by pouring new toner into the cartridge. Look for experience.

Are they a franchise? Beware of small franchises that come and go with the season. Many do not have the necessary technical background to properly do the job, and are merely following a manual. Sometimes the main company produces a quality product, but its franchises may not follow in its footsteps.

Do they have an engineer on staff? More specifically, do they have expertise? Who developed the technology that they are using? Is that person still with the company? Are they committed to ongoing research to improve services and develop new products?

Do they do complete remanufacturing? Beware of the infamous "Drill & Fill" operations! There is a lot more to remanufacturing than just replacing the toner. The complete process of remanufacturing is very involved, and requires the use of specialized equipment for completely cleaning the cartridges and drums of old toner residues. Components must be adjusted, drums polished, and parts lubricated. Beware of any company that does not remanufacture the fuser wand. It is critical to the proper operation of your machine, and must be properly lubricated.

Beware of companies that will send you the toner to "pour in" the cartridge yourself, or who sell cartridges that have been modified for "Drill & Fill" and are provided with bottles of toner and wands. These operations do not provide the critical servicing of the cartridge. You can usually tell if a cartridge has been drilled, because there will be a new hole in the top or the side of the cartridge that has been plugged or taped over. The correct way to remanufacture a cartridge is to completely disassemble the cartridge, clean the waste reservoir, polish the drum, clean and seal the toner reservoir, and fill it with new toner. A good remanufacturer will also make sure that all of the gaps between parts are properly spaced so you get the best results. If any of these steps are omitted, the cartridge may not work properly.

Do they use high-quality toner? Some rechargers use toners designed for copiers in their laser printer cartridges. While these may work, they are unlikely to produce the quality that can be achieved through proper remanufacturing. Look for toner quality at least as good as the original, and preferably better. The best remanufacturers offer high-quality toners that will produce better blacks than the original toners, especially in the Series I laser printers, which produce notoriously poor grays.

Do they seal the cartridge properly? If the cartridge is not properly sealed during the remanufacturing process, it will leak during shipping. Be suspect of companies that insist on hand-delivering their product. They may not have mastered the technology. A proper seal makes shipping possible, so you are not limited to the local kid down the block. The top remanufacturers do business throughout the country, and are not limited by geography.

Do they test the cartridges? Every cartridge should be tested in the appropriate laser printer or photocopier before it leaves the plant. This assures that the cartridge you get will be of the highest quality. Be suspicious of any recharger who does not even own a laser printer. They are unlikely to be able to provide quality products.

Do they offer new drums? There are new drums available for most cartridge models, which are better than the drums provided in the original cartridge. A premium-quality drum will have a harder surface that is more resistant to wear, and thus last longer and give better images. If a new drum is available for your cartridge, then get it. You'll be happier in the long run. Beware of coated drums. These are original drums that have been coated with some substance. We have found that these coatings come off very quickly, and do not improve the print quality at all.

Are their cartridges fully guaranteed? They should guarantee their work. If they don't, then avoid them like the plague. The last thing you need is a fly-by-night outfit. All of the major companies do guarantee their products. This relates to how long they've been in business. Their guarantee won't be much good if they go out of business next week.

Service and professionalism? Are they doing it in their basement on weekends, or is this their primary business? Look for companies that focus on remanufacturing. You don't want someone who does it as a hobby; you want professional results. Ask for their literature. Does it look professional? Are they friendly on the phone? Are their sales people well informed and helpful? Can they provide technical support in case you call with questions? Are they knowledgeable about their products? Beware of those who cannot be reached during normal business hours. This situation bespeaks a part-time job.

Packaging? The cartridges that come back to you should be properly repackaged in a new foil bag and box. The wand should be sealed in plastic tubing, to prevent the fuser oil from dissipating. Good packaging is a sign of a company that is proud of its products.

Zip It Up...

Many quality remanufactured cartridges, including those from BlackLightning, are packaged in resealable zip-lock foil bags. These bags allow you to reseal your cartridge for storage, and can be cleaned for reuse when returned to the manufacturer. This will help assure you of better quality, especially with cartridges that you periodically remove from the printer before they are empty, such as colors and transfer cartridges. Don't throw them away! Send them back! Reduce, reuse, recycle!

While most companies do refill the cartridge with at least as much toner as the original manufacturer, the total amount will vary. In addition, some companies offer options, including color refills and special graphics-quality toner. These options may increase the price of the product, especially for expensive color toners.

Pooling refers to the practice of exchanging your cartridge for another. With a high quality remanufacturer, pooling will not matter because all the cartridges are high quality, fully tested, and guaranteed. Pooling helps the remanufacturer keep costs down, and thus keep your price lower.

Most laser printer users have, or should have, more than one cartridge. We tend to recommend three cartridges per printer. Thus one is in the mail, one is being used, and the other is on the shelf, waiting to be used. Because of this, the turnaround time does not tend to be an issue, as long as it is not longer than a couple of weeks.

Don't Throw Away Your Used Cartridges!

Even if you don't have your toner cartridges remanufactured, you can still save money. Buy your new cartridges, and sell your empty toner cartridges to a remanufacturer.

Price is not a primary concern. As noted above, the price varies considerably. At the bottom of the range, you may want to be suspicious of the quality of the product; at the top of the range, you are not making significant savings over a new cartridge. You will save more money (in product quality and reliability) by going with a reputable firm, rather than going for the lowest price. With a good remanufacturer, you will experience significant savings. Don't skimp on the quality and gain a whole new headache.

In summary, look for a remanufacturer who has been in business for a while; has a good reputation; knows what it is they are doing; does a complete job; is friendly, helpful, and knowledgeable; uses quality toners; tests all products before they leave the plant; and is professional. By following these rules, you will probably find a firm that meets your needs, with quality at a savings.

Cartridge Shelf Life

Most cartridges have a shelf life of over one year. Store them in a clean dry area between 0°F and 95°F, in their original sealed packaging. Do not store cartridges where the humidity is high, or where there are abrupt temperature changes. Avoid exposure to corrosive fumes such as ammonia, benzene, and salt air.

Keep Me in the Dark

When storing cartridges, remember to keep them in a dry, dark, temperature-controlled area, preferably in the original packaging. The cartridge drums are photosensitive, and risk overexposure if the cartridge is not stored properly.

CARtridge Pooling

If you only go through one cartridge every year or two, you may run into difficulty with backup cartridges that sit on your shelf and age years beyond their usefulness before you ever get a chance to use them. Consider cartridge pooling with a local friend who has a similar machine. The backup cartridges will be put to use in good time, and there will still be a backup for both of you.

Making Toner

The Manufacturing Process

By Tom Durgin of BlackLightning, with assistance from Pat Bell of ITA

From *The Flash* Volume 3, Issue 1

Toner is the powdery ink that seems to magically move from the cartridge to the printed page. Surprisingly, its basic ingredients are actually common materials. The method of processing and delicate mixing, on the other hand, are very exacting and not so common at all.

Monocomponent toners are "premixed," while bicomponent toners go through a final mixing in the laser printer or copier itself. The popular Canon engines, found in a wide array of copiers and laser printers, are self-contained, using monocomponent toners. Ricoh machines, on the other hand, use bicomponent toners. The descriptions in this article are based on monocomponent toners.

Toner consists of a few basic ingredients, some synthesized, others natural. Toner is 35% to 40% iron oxide (rust) designed for specific magnetic properties. It is this magnetic charge and static charge that cause the toner to move from the hopper of the cartridge to the developer roller, to the cartridge drum, and onto the paper. The toner's primary ingredient, plastic (styrene or a styrene/acrylic blend), melts readily when exposed to heat. Thus, once on the paper, the powdery toner is melted into place as it passes through the heat and pressure of the printer's fuser rollers.

Some of the other materials added are used to help adjust these crucial characteristics. A type of sand (silica) acts as a free-flow agent, keeping the toner from clumping. A charge dye is used to adjust the static charge of the toner. Wax aids the dispersal of the toner as it melts with the heat of the fuser rollers.

For consistency, toner is made either in very large batches, the way paint companies make a batch of a particular color, or in a continuous process. In a large-batch process, variations in the toner may occur from batch to batch, but good quality control minimizes these differences. In the continuous process, changes occur only over a long period of time, and are also negligible.

One common method used to process toner involves heat and cold (Figure 1). The melt mixer has two large steel rollers. One is heated by steam, the other is cold. On the smaller scale, used by specialty toner manufacturers, the process of melt mixing bears a remarkable resemblance to making taffy. The rollers turn in opposite directions to each other. As the powdered plastic is introduced, it goes around the heated roller and folds back into itself. The cold of the other roller prevents the resin from sticking. Gradually, a long roll of resin is formed in the nip, where rust, sand, wax, carbon black, and dye are added. When the roll has reached the correct proportions, it is cut from the steel rollers with a large blade. After being folded like a flag, the sheet is reintroduced to the melt mixer. The rolling and folding process is repeated several times. When the roll is removed from the machine for the last time, it is allowed to harden.

This hard sheet of toner is coarsly ground into small pellets, and jet-milled to a fine powder. Care must be taken not to grind the particles too uniformly; toner actually works best when the particles fall within a certain range of sizes. If the proportion of sizes in the toner is off, it will result in spotting and smudging on the print. The range of proper particle size is from 5 to 15 microns, so fine that the dry toner looks and acts much like a fluid.

Classification of the toner uses centrifugal force to separate the samples by size. The test toner is then processed through a machine normally employed to count blood cells, called a "Coulter Counter." The results of these tests are plotted and graphed. A "postblend" may be done to fine-tune the mix with additives that control charge, flow, and other properties.

FIGURE 1
Toner manufacturing

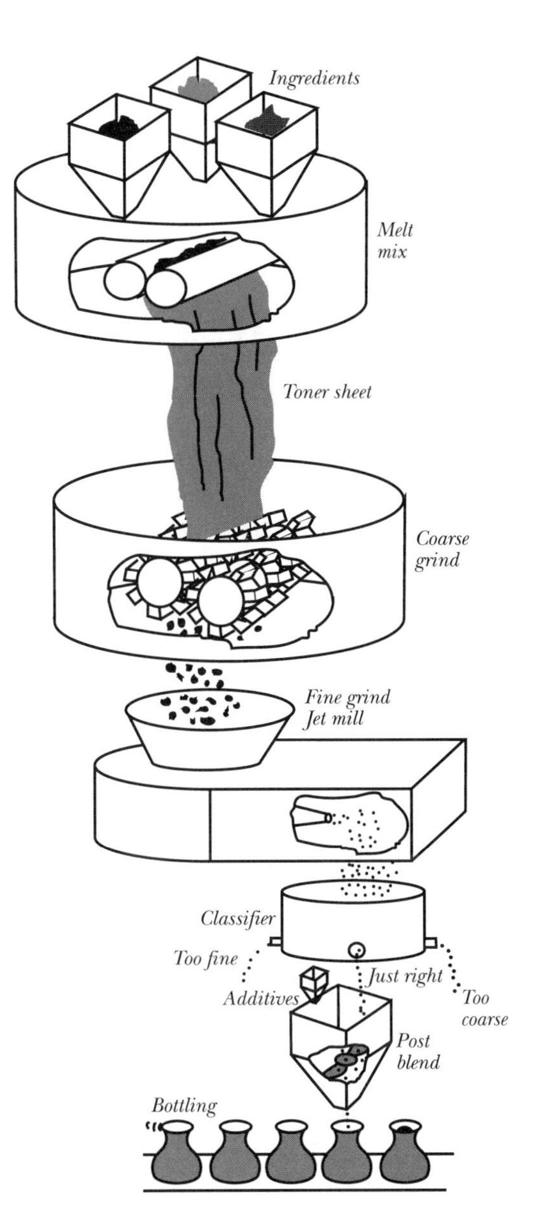

Color Toner

As for the color, both the plastic and rust turn black during the manufacturing process. Carbon black is added to further deepen the blackness of the print. Making any other color is difficult, because the normal black of the toner must be masked. Unfortunately, masking affects more then just the print color.

The magnetic charge of the black iron oxide may be affected by the addition of pigment or dye. Similarly, dye may change the melting point of the plastic, which must stay within the range of requirements of the printer's fuser roller assembly. The result is that monocomponent color toners are very hard to produce. This means that the color choices are limited for the popular laser printers based on the Canon printer engines.

Toner is a mixture of unremarkable, ordinary substances. Yet when mixed properly, it is capable of making the most sophisticated printers and copiers produce remarkable results.

Emerald Drums

Send Us Your Tired,
Your Poor, Your Weary...

By T. Fable

From *The Flash,* Volum 4, Issue 1

Technology keeps striding forward, even in the inner sanctums of toner cartridge manufacturers. Whereas the original toner cartridges from Canon, Apple, H-P, and the like (including most remanufacturers) use acrylic drums that are prone to scratching and breakdown over time, you can now get cartridges with polycarbonate drums that resist wear and are more reliable. At BlackLightning we call our polycarbonate drum the "Emerald Drum."

Drum wear has been the Achilles' heel of the remanufacturing industry for years. BIS/CAP International, a market research firm, stated in 1988 that the short lifespan of the EP-S cartridge's drum was the most significant hurdle facing remanufacturers, preventing many consumers from remanufacturing their cartridges. The same difficulty applies to new EP-L and PCmini cartridges.

But It Was Fine When I Last Used It....

During the final stages of cartridge use, drum defects may not be visible on the output. Since the toner is not as dense at that point, scratches and dents may not show up. However, after the drum is polished and the cartridge refilled with new toner, any flaws will become apparent. For optimum performance, cartridge drums must be replaced when they show signs of wear, including scratches, nicks and exposure lines.

While the older Series I (CX) or EP cartridges have a large 2.44-inch-diameter drum that typically lasts for four to six cycles, a careful failure analysis shows that the newer cartridges (1.25-inch diameter) often wear out after one or two recyclings. Early failure of the drum causes lines, stray marks, and lower-quality output, dampening the spirits of many a recharger and user alike.

To get beyond the two-cycle limit of the newer cartridges, a new, harder drum was needed. For over two years, researchers worked on developing just such a drum for the rapidly expanding recharging industry. In 1989, preliminary samples were distributed. We performed exhaustive tests and analysis of the candidates, first inhouse and then in the field. The results are nothing short of remarkable. One Emerald Drum was reused 46 (!) times, with no decrease in copy quality! It's a laser printer user's dream.

Until recently the drums have been very expensive, costing upwards of $60 each. The initial cost kept Emerald Drums out of the hands of average laser printer user, leaving it to those with foresight to see that the actual cost is much lower, because the drum does not wear as rapidly. But now a standard remanufactured Emerald cartridge costs less than a new cartridge.

I See the Light!

Drum overexposre is caused by excess light. Even with the protective cartridge drum door, if it is left in a window or in bright light it may become overexposed. This results in a dark brown or dark orange line, often spanning the length of the cartridge drum. If overexposed, the photosensitive drum does not respond properly to the laser beam of the printer, resulting in poor-quality output. Protect your drum by storing it in its box or in a dark room when it is not in the laser printer.

Prime Time

After printing five to 10 solid black pages, your cartridge will print blacker blacks. This process primes the cartridge drum, causing a stronger, more even layer of static electrical charge to be laid out on the surface of the drum, thus pulling more toner from the reservoir and allowing the cartridge to print at its very best.

Copy Counts

The Life of a Toner Cartridge

By John Jeffries of BlackLightning

The number of copies you get from a cartridge will vary. Print coverage, machine density setting, drum type, toner type, cartridge tabs, and gapping of the cartridge will all affect copy count. The most significant factor is print coverage—how much toner you use up on each page. The more print coverage, the more toner used, and thus the fewer pages per cartridge.

The standardized definition of coverage, used by Hewlett-Packard, Canon, Apple, and other manufacturers for copy counts, describes a short letter which covers less than one third of the page with text at 5-percent coverage. An EP-S cartridge normally prints out approximately 4,000 to 5,000 short letters. If printing only pages of solid text, it will run approximately 800 pages. If you print solid-black pages, you'll get even fewer copies from a cartridge.

On the other hand, one of our customers got 12,000 copies from his EP-S cartridge while working on a directory which had very little text per page. Changing the density dial on the printer to a lighter setting will conserve toner and make the cartridge last longer.

The drum type can also make a difference. We have experienced that after ten cycles, the Emerald polycarbonate drum produces more copies per fill. This is not something that we guarantee, but it is something that we have observed. Setting the tabs of the cartridge to a darker setting will cause the toner to be applied more densely, using more toner per page.

Some cartridges have "page counters" or "indicators" on them that change from green to yellow to red, as the cartridge is used up. These count rotations of the drum, and may or may not reflect how much toner remains. Other printers give a signal saying "toner low." Again, this may be right or not. Both of these will vary, depending on the toner used per page and even the humidity for the latter type of indicator. The best bet? Use the cartridge until it no longer gives good print. This may well be after the indicator says you're out.

Toners used by different manufacturers will provide varying yields per gram, and the gapping of the cartridge done by the technicians affects the number of copies produced. Furthermore, with graphic toners you trade yield for darker copy. The copy counts listed are approximations. Like the mileage on your car, it depends on how you drive it, where you drive it, and how you load it up.

Remember, your mileage may vary.

TABLE 1 **Copy counts chart**

	EP	**EP-S**	**EP-L**	**EP-F**	**PC**	**A15**	**A30**	**MP-N**	**MP-P**	**R6000**	**R4080**	**Kyocera**
New	3,000	4,500	3,000		1,500	1,500	3,000			1,500	4,500	3,000
Standard	3,000	4,500	3,000	3,000	1,500	1,500	3,000	3,000	3,000			
Long Life	4,500			4,500	2,000	2,000	4,500	4,500				
Text	4,500	4,500										
Graphic	4,500	4,500										
Color	2,000	2,000										

You Know You're Near the End When...

the waste reservoir in your cartridge begins to overflow. There is a waste reservoir in every CX, SX, and LX toner cartridge that catches any toner that does not manage to get onto the paper. This overflowing will often cause streaking on the page. This streaking may be an indication that the cartridge has almost no toner left in it. Try cleaning the corona wires in the cartridge, and try rocking the cartridge to get a little more use out of it.

A One-Sided Affair

If you do a large amount of printing on one part of the page (many copies of a dark graphic or text, just along one edge) then you may fill up the waste reservoir in line with the image, causing it to overflow. This can produce ghosting, smearing, and spilled toner in the machine. The solution is to gently rock the cartridge at approximately a 45-degree angle to redistribute the toner. Clean the corona wires in the cartridge after you do this, to remove any toner that may have spilled internally.

Here are some tricks to avoid jams.

- Keep the paper dry, and wrapped or boxed until it is to be used. Consider using a sealed, airtight, vaporproof container (such as a Tupperware or Rubbermaid storage bin) until you put it into the printer. Adding some desiccant packages will also help.
- Only put as much paper into the printer as you need right then.
- The worst jamming and wrinkling of the paper is caused by the leading edge of the paper curling. Bending it flat or slightly down after the first pass helps a lot.
- Between passes, store the print job on a flat surface, face down beneath a heavy flat object, to minimize the curl and restore flatness. I use a piece of granite.
- Keep your laser printer clean and in good repair.
- Change the ozone filter on time.
- Keep the machine cool, and in as dust-free an environment as possible.

One trick I use when duplex-printing a document from a program that won't let me print just the odd pages is to print the whole document, and then switch every other page, put the pages back in the paper tray other side up, and print again. Now page one will print on the back of page two, two on the back of one, three on the back of...

Used Laser Printers

Getting a Good Deal

By Walter and John Jeffries of BlackLightning

From *The Flash,* Volume 4, Issue 1

One of the questions most frequently asked by *Flash* readers is, "What about used laser printers?" We have found the laser printers based on the Canon CX (EP) and SX (EP-S) printer engines to be very rugged machines. These include the Hewlett-Packard LaserJets, Apple LaserWriters, and printers from many other manufacturers. We liken these robust machines to the Model As of years gone by. They have enough features to make a very serviceable machine, yet are so well built they last way beyond their expected lifetimes. We know of numerous LaserWriter Plus machines that have printed over a million copies, and are still running strong. Our own LaserWriter, purchased used years ago, has printed over 500,000 copies and we expect to keep using it for years to come.

For a good source of used laser printers, check out your local university or college. Large corporations are another good bet. They regularly upgrade their equipment, and sometimes sell off old machines at a real bargain. Several years ago I bought three Apple LaserWriter Plus printers for $1,400 each. This was when they were typically selling for over $2,500 from the used equipment houses found listed in the backs of computer magazines such as *PC Magazine* and *MacUser,* as well as the yellow pages. Scout around; you may find a great deal.

Of course, you have to be careful not to get burned. If you are not familiar with laser printers, take along a friend who is. Try to get some form of written guarantee (30 days is reasonable) from the seller. They may not be willing, but it is worth asking. At least run the machine and print out some copy, so you know it basically works.

Which machine should you get? There are three basic factors to consider:

- Printer engine type and manufacturer
- PostScript or non-PostScript
- Interfacing to your computer

Printer Engine

The printer engine is not critical, although it is simpler to have just one type in an organization because the cartridges and other parts are not interchangeable. The Canon printer engines are the best, and have the added advantage of being usable with transfer toner. The Canon engines come in several types, the most popular being the CX and SX. The advantage of the SX over the CX is that you can use the long-lasting Emerald drum for your cartridges, and some people claim to get darker blacks with the standard toner. We have not discussed other printer engines because most of our experience has been with the Canon CX- and SX-based laser printers. Others, such as the Kyocera, Ricoh, and Canon LX, may or may not be good buys in used machines.

PostScript

If you are going to be working with graphics or precise type, then you will probably want PostScript. PostScript is a printer language that lets you do the greatest graphics and text this side of Hell*, and some of their machines use it, too. Its biggest feature is that fonts and graphics can be enlarged or shrunk (scaled) without getting the jaggies.

*Linotype-Hell Co. is a producer of high-end imagesetters.

However, you can't take advantage of Postscript unless the software on your computer also works with Postscript. Check your software's reference manual if you are unsure. All Macintosh and Windows programs work with PostScript; most DOS programs work with PostScript, including all the major word processors, spreadsheets, and page-layout packages.

Interface

If your computer won't talk to the laser printer, then you've got one very expensive paperweight. Fortunately, most computers now connect to most printers. At worst, you might need a special cable or software driver. The interface consists of two levels: the hardware (communications port and cables), and the software drivers. Macintoshes, in general, only work with printers that have a serial or Apple/LocalTalk port. PC computers typically accept both serial and parallel ports (often called a Centronics port).

The software to control the printer from your machine must know how to talk to the software on your printer. Ask the person selling the printer about this. If they can't help you, try calling the printer manufacturer. The number should be in the front of the manual.

There are a lot of printer drivers that have been written for the PC, but until a few years ago Macintosh owners had very few choices. They could use an Apple laser printer, or they could use an Apple dot-matrix printer. The printer-driver software that was available for the Macintosh was not very well implemented. Things have changed significantly, though; there are now several very good packages available that dramatically expand the printer choices for Macintosh users.

One package that we have used is the JetLink Express from GDT Softworks, Inc. The parallel version we reviewed is readily available in the USA for $249 directly, or for $179 from MacConnection (the serial version costs $159 and $89, respectively). JetLink Express comes with a serial-to-parallel translation cable; a small wall transformer to power it; four disks containing seventeen printer drivers for the Canon BubbleJet, Canon LBP Laser, H-P LaserJet, H-P DeskJet, and H-P DeskWriter printers; and a thick manual.

I ignored the manual, inserted the H-P drivers disk, and plugged my H-P Personal LaserJet IIP into my Outbound Wallaby (a Macintosh-compatible portable). No luck. After a trip to the manual, I called tech support. They were very helpful, and referred me to the HyperCard setup file on the disk. This little program has step-by-step instructions on quickly getting started with all the popular machines. Moments later, my Mac and the H-P IIP were happily chatting away, spewing out graphics and text. I tested JetLink Express with a wide variety of software, including HyperCard, MS-Word 4.0, MS-Excel, Sys1, Paradise Market II, MacPaint, DeskPaint, MacDraw, and many others. The only program I couldn't use was Adobe Illustrator, which requires a PostScript printer.

The software was very intuitive and Mac-like. I'd give the software an A+, and the manual was topnotch as well. Things I like to see in a manual are a well-thought-out table of contents, system requirements, a quick-start section, detailed descriptions of all the program's workings, a troubleshooting section, a glossary for unfamiliar terms, and (of course) a comprehensive index. JetLink Express gets thumbs up in all areas. The manual's convenient 5.5-by-8-inch size, well-laid-out screen shots, and carefully chosen fonts make for easy reading. In addition, it includes an easy-to-use compatibility chart for both printers and software. GDT appears to have tested their software on 32 different printers, and it should work with many others that have Canon or H-P printer emulation. Lastly, the manual had an Application Notes section with many of the little tricks to use with 29 major software packages—a real time saver. If you need to connect to a Hewlett-Packard or Canon printer, check out the JetLink Express. We're impressed! GDT Softworks, Inc., 4664 Lougheed Highway, #188, Burnaby, BC, Canada V5C 6B7. Telephone 604-291-9121. Fax 604-291-9689.

Your Best Face

Getting Quality Copy with Your Laser Printer or Copier

By Holly Blumenthal of BlackLightning
From *The Flash,* Volume 2, Issue 1

With some careful observation and a bit of detective work, it's often possible to pinpoint the cause of poor output that occasionally emerges from a photocopier or laser printer. This article describes some of the most common and most likely causes of poor print quality.

When trying to determine the cause of a mysterious smudge it is helpful to have a basic understanding of how a laser printer or photocopier works. The paper is picked up from the paper tray and moved through the machine by a series of rollers along the paper path, past the cartridge's drum, and through the fuser assembly. In the developing process, the corona wire in the cartridge prepares the cartridge drum, providing a "clean slate." The laser beam in the machine uses an electromagnetic charge to apply the image to the surface of the drum. The toner is then attracted to these charged areas. At this point, the print can actually be seen on the drum surface. Meanwhile, the paper passes over the corona wire in the machine, which charges the sheet so the toner is attracted from the drum to the paper. With the powdery toner on the surface of the paper, the sheet passes through the fuser assembly, where heat and pressure melt the toner.

First Step: Change the Cartridge

When a problem does emerge from a printer, one fast and easy trick to narrow down the list of suspects is to change the cartridge in the machine and see if the problem disappears. If the troublesome markings are still there, then an element in the printer or computer is the most probable source of difficulty. If the problem disappears when another cartridge is used, then an element within the cartridge is the likely cause.

Keep in mind that this is not an absolute test. There may be something misaligned in the printer which is wearing on the cartridge drums and ruining each cartridge as it is used, or it may be an interaction between printer and cartridge, or an environmental problem. A new cartridge might not reveal the problem initially, but after being used for a time, this same (now not-so-new) cartridge will begin to print poorly. If you have difficulty with several cartridges in a row, it is highly recommended that the printer be cleaned or serviced. For more information on providing basic maintenance and service for your machine, see the CX, LX, and SX maintenance chapters at the beginning of the book..

The troubleshooting table on the following pages describes a range of problem markings, possible causes, and appropriate solutions. The descriptions are divided into three basic sections, labeled white, gray, and black. Diagrams are provided to facilitate quick and easy reference. Following this section is a more detailed discussion of many of the solutions listed. A quick comparison of these pictures with your machine's output may save a costly service call.

Duplex Living

Although they are available, most of us don't have double-sided laser printers or copiers. But we still need to print in duplex mode from time to time, whether it is to save paper or to produce a publication that has a more professional look. A single-sided laser printer or copier can print double-sided, but there are a couple of problems to be overcome.

The first is paper jams. When the paper goes through multiple times, it can begin to curl and is more likely to jam.

Fortunately, we rarely experience jams on the second pass. Usually it is the fourth, fifth, or sixth pass where jams become markedly more frequent. This isn't usually a problem unless you're doing multiple colors and multiple sides, for example with a short-run brochure. Jams are worse in humid weather.

You now have many clues which can be helpful in determining the most likely cause of your printing difficulty. When the printer begins to spit out poor quality prints, take a few samples and make some careful observations. Match your output with the descriptions above, and narrow down the list of possible culprits. May your magnifying glass shine, and may your investigations be quickly resolved.

TABLE 1 **Troubleshooting**

White

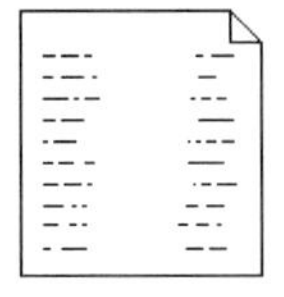

Middle or side section of the page is not printing

Cause	Remedy
• Dam is not pulled	Pull the dam.
• Toner packed to one side	Rock the cartridge.
• Empty cartridge	Insert a new cartridge.
• Dirty or broken cartridge corona wire (copiers only)	Clean the corona wire or return for repair. See appropriate maintenance chapter.
• Missing spring in cartridge	Return it for repair.

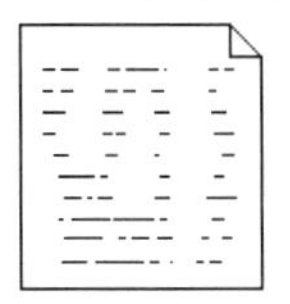

Vertical white streaks

Cause	Remedy
• Dirty machine corona wire	Clean machine corona wire with alcohol and Q-tip.
• Damp paper	Open new ream of paper and/or get a dehumidifier. Store paper in an airtigh container, possibly with desiccants.

TABLE 1 **Troubleshooting** (continued)

	Problem / Cause	Solution
	• Dirty cartridge corona wire (copiers only)	Clean the corona wire with the cartridge corona wire-cleaning tool. See appropriate maintenance chapter.
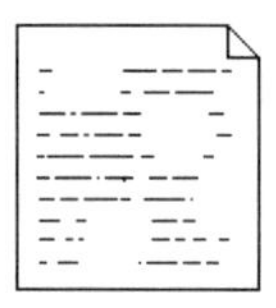	**White spot or blotch**	
	• Dirty machine corona wire	Clean machine corona wire with a Q-tip and alcohol.
	• Paper surface is wet	Use a different batch of paper, and dehumidify the room.
Gray		
	**Light print**	
	• Density setting on "light"	Adjust the density dial.
	• Cartridge break-in period	Print 50-100 pages.
	• Cartridge tabs	Adjust cartridge tabs.
	• Missing spring in cartridge	Return it for repair.
	• Cartridge gapped too tightly	Return it for repair.
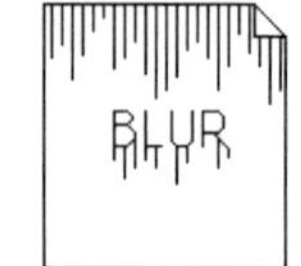	**Thin gray streaking from top of page, or blur from bottom of print**	
	• Dirty fuser wand	Clean or change fuser wand.
	• Paper hitting cartridge drum because it is too thick or wavy	Use different paper or dry and flatten paper before using.
	Ghosting of previously printed images, typically from same page	
	• Lower fuser roller may be dirty	Run 20 or more blank pages.
	• Upper fuser roller not cleaned by fuser wand felt	Install or replace felt.
	• Cartridge wiper blade is improperly positioned, or old	Return for repair or replacement.

TABLE 1 **Troubleshooting** (continued)

	Problem / Cause	Solution
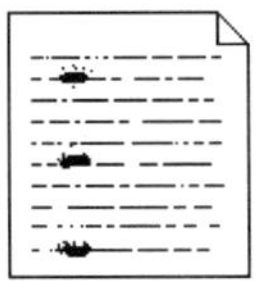	**Regular spots of gray or black background**	
	• Humidity too high or too low (graphic toner is particularly sensitive)	Adjust humidity of room.
	• Static charge build up	Ground the cartridge by touching the metal contacts to a grounding wire or grounded water pipe.
	Consistent, evenly spaced "shadow spot," usually three or four times per page	
	• Drum out of round (EP-S usually), usually caused by dropping or knocking the cartridge, or by cartridge wear	Return it for repair.
	• Cartridge not properly seated in printer. Printer may be misaligned.	Try removing cartridge and reinserting it.
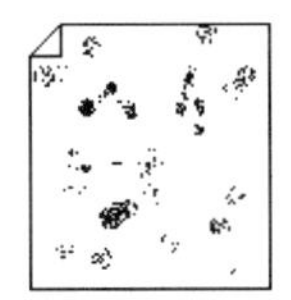	**Marks on back of page, consistent location**	
	• Dirty pickup rollers in the machine	Clean the rollers, located over paper tray, with acetone.
	• Dirty lower fuser roller	Clean by running 20 or more blank pages. Replace fuser wand felt, or run fewer double-sided pages. Use appropriate felts.
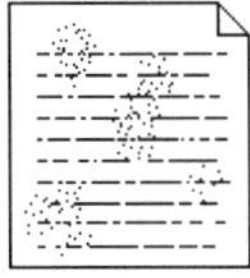	**Smudging or backgrounding**	
	• Humidity too high or too low (graphic and transfer toner are particularly sensitive)	Lower humidity.
	• Dirty fuser wand	Replace the wand or felt.
	• Cartridge gapped too loosely	Return it for repair.
	• Dirty magnetic bar (evenly spaced three or four times down the page)	Return it for repair.

TABLE 1 **Troubleshooting** (continued)

Spotty gray vertical line

Cause	Remedy
• Dirty separation belt, line on right edge of paper (Series I laser printers and PC 10 through PC 25 copiers only)	Clean the separation belt.
• Loose toner from handling	Print 10-20 pages.
• Fuser roller wear (line will match location of wear on the fuser roller)	Replace the fuser roller.
• Humidity too high or low	Adjust the room's humidity.
• Dirty paper feed rollers	Clean the rollers.

Black

Blurred print, or print too dark

Cause	Remedy
• Density setting on dark	Adjust the density dial.
• Cartridge tabs	Adjust cartridge tabs.
• Printer overheating	Make space around machine. Vacuum the printer's vents. Change the ozone filter in the printer.
• Low humidity	Adjust room humidity. Use a higher-grade paper.

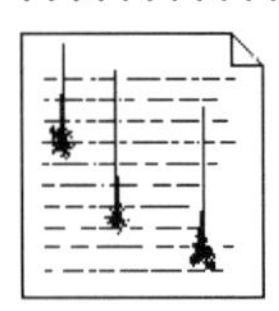

Solid black vertical lines like paint drips

Cause	Remedy
• Dirty cartridge corona wire (EP, EP-S, EP-F, and MP-N cartridges only)	Clean cartridge corona wire with the cartridge corona wire cleaning tool. See appropriate maintenance chapter. Do not use this tool's felt for cleaning in the machine, or you may dirty it. Vacuum tool, if it is dirty.

TABLE 1 **Troubleshooting** (continued)

	Problem / Cause	Remedy
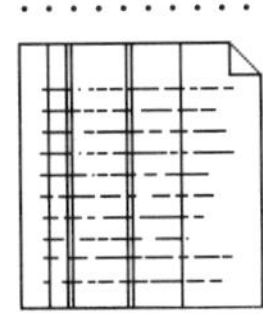	**Consistent, vertical, thin, black lines**	
	• Dirt on fuser assembly claws	Clean the claws under fuser assembly lid.
	• Fuser assembly wear	Replace the fuser roller.
	• Scratch on cartridge drum	Return for replacement of drum.
	• Hair making contact between drum and cartridge corona wire	Clean cartridge corona wire.
	Pin marks, evenly spaced in consistent vertical line	
	• Programmed into document (i.e., a period or a dash)	Correct document.
	• Pin mark on the cartridge drum (usually appears three or four times per page)	Return for drum replacement.
	**Random spotting**	
	• Dirty glass (copiers only)	Clean the glass.
	• Dirty fuser roller wand	Clean or replace fuser roller wand.
	• Marks originally on the paper being used	Use paper most appropriate for your needs.
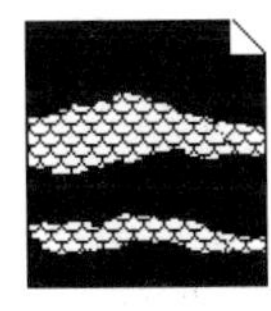	**Black pages and "lilypad" pattern**	
	• PC cartridge not inserted all the way into the machine	Push cartridge snugly into the machine so that complete electrical contact is made.
	• Broken cartridge corona wire	Return for repair.
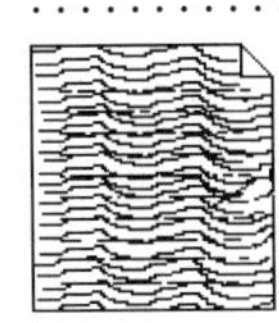	**Wavy background and halftones**	
	• Too high a charge on toner	Return for repair.

Clear as Black

Getting the Best Blacks from Your Laser Printer

By T. Fable of Blacklightning

There are several methods you can use to improve the black output from your laser printer—from special toner to spray-on enamel.

Density Control

The darkness of the print can be adjusted with the density control located in the machine. It is also important to remember that a cartridge may not print at its best capacity until the break-in period is complete. On remanufactured cartridges, the break-in period is typically 50 to 100 sheets. For new cartridges, it may be as many as 100 to 200 pages. Printing five to 10 solid black pages, or 15 to 30 blank pages and a few black pages, can expedite this process.

Cartridge Tabs

The tabs on the end of the cartridge also affect density. If you are getting light copy from a full PC or EP toner cartridge, try removing one or both tabs (see Figure 1). If you are getting overly dark copy or backgrounding, try printing with both tabs

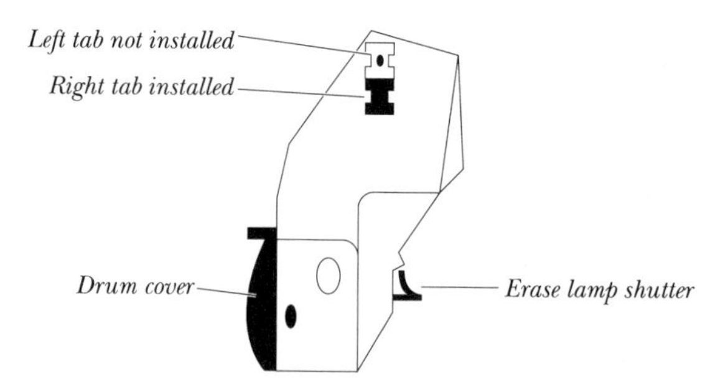

FIGURE 1
Cartridge tabs

inserted. Extra tabs can be found in unused or empty cartridges, or made with cardboard and scotch tape. EP-S cartridges work the same, and must have at least one tab.

Graphics Toner

Special graphics toner like that available from many remanufacturers will give you the darkest black possible from a toner cartridge. When using the cartridge for special desktop publishing documents containing large black areas such as oversized type or boxes, running several pages of solid black copy before printing will produce the blackest copy possible. This primes the drum up so that it will produce at its absolute best.

No toner cartridge will provide an absolute black. If an absolute black is needed, use an imagesetter at a local service bureau (or see the next chapter, *Negative Images*). Imagesetters print with a laser beam onto photographic film at very high resolutions. You might also want to investigate high-resolution toner for the newer laser printers.

Spray-On Enamel

Use spray-on enamel such as Krylong Crystal Clear acrylic spray to make large black areas on laser-printed artwork a uniform, solid black. Use a fast-drying clear spray enamel or acrylic—available at art supply stores, but a lot cheaper at your local hardware or discount store (typical price $1.49).

Once you've printed your art, find a well-ventilated area. Spray a quick light coating of enamel onto the page. Be aware that fine detail may "melt" a little. Try to avoid spraying these areas, either by directing the spray or by cutting a mask that exposes the target area. Sprayed blacks darken, solidify noticeably, and become glossier and more scratch resistant. Use the results as camera-ready copy for printing, or as final artwork for presentations, signs, etc.

Blacker with Dot Gain

When you are making laser prints of something that will then be photocopied or offset printed, you do not need to have completely solid blacks. Because the inks in offset printing are pressed onto the page, they spread a little. Also, the technology commonly used to make the printing plates cannot support more than about a 150-line-per-inch screen; thus the white lines you see in the solid black areas will fill in. This is called *dot gain*, and results in dark things getting darker. This is why a photograph in your newsletter may print too dark.

You can compensate for this by lightening the image when you print it on your laser printer by using software like PhotoStyler or Photoshop. You can take advantage of this effect, because it will darken up solid areas on your output that should be absolutely black, but aren't.

High Yield versus Dense Blacks

There is a trade-off between getting many pages from a cartridge, and getting the darkest-possible graphics. Basically, you need to put more toner on the page to get darker graphics, so the toner gets used up faster. You can adjust the density with the density control on your laser printer and the tabs on EP, PC, and EP-S toner cartridges.

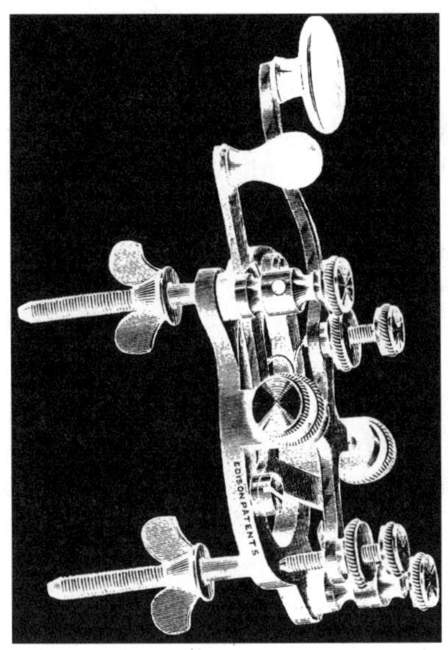

Negative Images

Making Film Negatives and Positives

By Rodney Grantham
of Grantham Printing

From *The Flash,* Volume 4, Issue 3

Have you ever needed a film negative or positive? Say, for shooting silkscreens or making rubber stamps? Or just as a transparency for overhead projection? Maybe you sought out a printer with a process camera. Or you tried to do it on your laser printer or copier, but couldn't get a dark enough image.

If you have a laser printer, or even a copier, there is no reason you can't make them yourself. Transparency film for laser printers, intended for use in overhead projectors, is available in most office supply stores. (Be sure to get the kind for laser printers, so it won't melt in the printer's hot fuser roller assembly!)

This works, and you can easily print on it with your laser printer, but there is one big problem. The fly in the ointment is that there is just not enough density. Toner just isn't black enough to mask the light completely—so you get grays. The reason for this is twofold. First, the transparency doesn't have quite the proper static electric properties to take enough toner off the imaging drum. Second, when light is passing through the toner from behind, the toner is not dense enough, even with the densest graphic toners, to completely mask the transmitted light. This results in a gray image.

Two Negatives in One

The trick? Place two transparencies together in register with each other. When aligned, the additive effect of the two layers of toner becomes dark enough for almost any shoot. Pretty simple!

To make a negative, start by drawing a black rectangle to size. Use white letters or artwork in the box, so they appear clear on the transparency. To get two pieces of transparency film to register, draw small ovals on top and bottom, outside the black rectangle. It is much easier to align thin black lines on a clear background. If you cut the top transparency an inch smaller than the bottom, after printing, then they can easily be taped together. A line thickness for ovals should be very thin, .003 if possible. A sheet of white paper under the transparencies will make the task of aligning them easier.

Rodney Grantham is a graphic artist and printing services designer who has developed a lowcost machine for making rubber stamps from laser printer negatives. He can be reached at 218-773-0331.

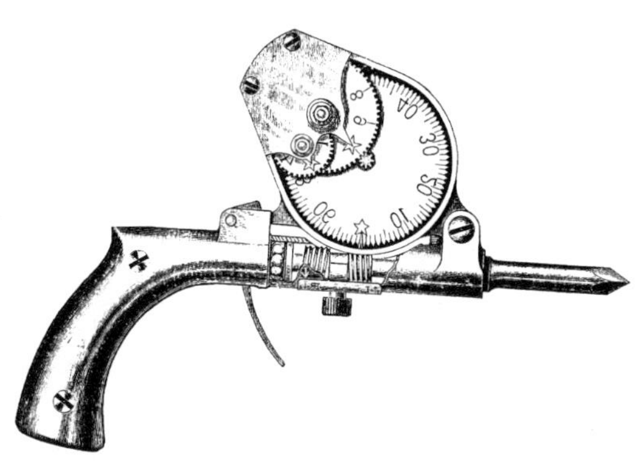

Paper Facts

Choosing the right paper can make all the difference

By Barbara Brooks of BlackLightning
From *The Flash*, Volume 1, Issue 2

When a customer asked for information on paper products, I thought it would be a piece of cake. After much research, I have realized that there's a lot more to paper than meets the eye.

This article is designed to help you make the best paper-buying decisions to serve your different laser printing and photocopying needs.

Be wary of extremely smooth or shiny papers, or of those that are highly textured, which are not laser printer compatible. Paper with special coatings may also give you problems. Letterhead that uses low-temperature dye or thermography may transfer onto the fuser roller, smear, and cause damage. Any pre-printed paper should have inks that can stand temperatures of 200 to 400 degrees for 0.2 seconds. Most papers printed with ink handle this fine; be careful with papers printed with thermography (like wedding invitations from a quick-print shop). If you're having stationary printed, tell the printer that you'll be using it in a laser printer or photocopier. Never try to send paper with surface characteristics or irregularities (such as staples, tabs, or wrinkles) through your machine. Carbonless paper or multipart forms won't work either. The paper path of a laser printer is tight, and the use of very light or very heavy paper can cause jamming.

Let's start with a glossary of paper terms you should be aware of when contacting sales people and dealers for the right paper for your machine.

Bond. This is 100-percent chemical wood pulp and/or cotton, with an ash content of less than 10 percent. Papers should not contain large amounts of clay or talc, which can damage the laser printer by abrasion.

Brightness. This describes the whiteness of the paper. A rating of 81 characterizes a basic paper for general inhouse use. Use it when economy is an important factor in your printing. A mid-range paper for a more professional, yet informal look, would have a brightness level of 84. A brightness rating of 86.5 will give that crisp, snappy look to more important documents.

Electrical Properties. This refers to the ability of a sheet of paper to hold or release an electrical charge. This is very important in any type of copying. The paper can't be too conductive, or the toner will not be attracted to it. On the other hand, if the sheet is too resistant, static cling and feeding and jamming problems will occur.

Moisture Content. You need to consider moisture content because of paper curl and static build-up. The wrong moisture levels can cause the laser printer to jam due to paper curl, static buildup, and wavy edges. Changes in electrical properties produce light spots. To ensure the proper moisture content, store paper in a clean, dry area that will not be subject to dampness. If high humidity is a problem in your geographical area, try storing the paper with a desiccant.

Opacity. A paper's opacity refers to the amount of show-through a paper has. High opacity is important when printing on both sides of the page.

Paper Curl. This is a critical factor for top machine performance. Paper should be basically unreactive and flat, although a certain amount of built-in curl is necessary to counteract the excessive heat and pressure of the fusing process.

Surface Characteristics. Smoother paper produces less wear on your laser printer, and a higher-quality image will result. The best finish is one that is free of dust and dirt, and isn't too abrasive. A certain amount of surface friction is necessary for satisfactory feeding. Card stock, a much heavier paper, has a rough surface and will not run through laser printers very well. Because of the roughness, toner may not stick evenly to this surface .

Always fan the paper before placing it in the tray. This allows air between each sheet, and helps eliminate blank pages and paper jams. Position paper in the tray as indicated on the ream label, or you may have difficulty in printing, regardless of paper quality.

The best-quality paper for use with laser printers has a clean surface, high opacity, and a brightness rating of about 84. A 70-pound smooth opaque such as Dello Opaque English Finish from James River Corporation, LaserPlus from Hammermill, or Simpson's Watermark will all work well.

The paper selected for most of the general printing at Black-Lightning is Hammermill Tidal DP, a high-quality paper at a bargain price. This paper is for use in xerographic copiers, offset duplicators, and laser printers. Tidal DP is lightweight, has a clean, smooth surface and a basic paper brightness of 81. This lightweight paper is ideal for use in multisheet mailings, but the opacity is not high enough for double-sided printing. Xerox 4024 and Canon NP are very similar papers.

Top-of-the-line paper for desktop publishing is specially designed for quality copies on the new high-resolution laser printers. This paper allows the user to produce camera-ready copy directly from the laser printer. The bottom side is specially treated to prevent wax bleed-through when waxing in preparation for paste-up. You might consider using James River's LaserUltra, Hammermill's LaserPrint or LaserPlus, or Laser-Bond by Boise Cascade for this purpose. One drawback of coated paper is the increased chance of the toner rubbing off. This can be prevented, however, by spraying the finished page with a thin coat of clear artist's fixative.

A wide variety of colored paper is available for use. Apply the same standards to colored paper as you would to any other paper. With a little experimentation and imagination, the results can be beautiful. Customers find that brown toner is elegant on cream-colored paper, as is blue toner on light blue paper.

Bicolored paper, such as QuickColor from Intergraphix and James River's Pro-Tech, is available to easily make brochures and pamphlets.

Raising a Point About Raised Print

Thermography (raised print), which is sometimes used on letterheads and business cards, will dirty your laser printer because the raised ink is a wax that will melt on your fuser rollers. It smudges on the page as it goes through the heat of the fuser assembly, and may also wear down the fuser roller. We suggest that you avoid using pages with raised print in the laser printer, and when considering the purchase of a special letterhead, first ask for samples to try. If you must use paper with raised print, plan on changing your fuser wands very often. This will help keep the printer clean, giving you better copy and extending the life of the fuser rollers.

Another product of interest is transparency film. Avery's transparencies for desktop laser printers are paper-backed for better feeding and easy proofing. 3M's Type-501 transparency film gives a very good performance, but because of the high fuser roller temperature, Type-503 will melt in your laser printer.

Laser Labels

If you want to make labels with your laster printer and have been warned not to, rest assured—there are labels suitable for your needs. Avery has introduced labels specifically designed for laser printers. They won't melt or peel, they go through the machine beautifully, and the adhesive adheres immediately. Dennison also manufactures laser printer labels that work in machines that have up to a 20-copy-per-minute capacity. After testing this product, however, we found that once the labels are put on an envelope or postcard, they have a tendency to peel off rather easily, and are more prone to jamming.

When printing envelopes, be aware that those with five layers in the middle (where the folds meet), may not pass very well through your printer. This is due to too many thicknesses of paper going through the printer at once. When trying to print on the front of the envelope, the type may come out light on one side and dark on the other. Envelopes that have only two thicknesses in the back are better for use with laser printers. We have found it is cheaper and more expedient to have our envelopes made up by an offset printer, and to use labels for addressing.

At BlackLightning we recycle everything from your cartridges to paper to soda cans. We thought some information on recycled paper would round out this column.

- EarthRight, a Vermont company which promotes recycling, stocks 70-percent recycled white bond paper, available by the ream or by the case. For more information, call 802-295-7734 or 802-649-1008.

- In Massachusetts, Recycled Paper Company has 20-pound bond that has an 81 brightness rating. This paper is more than 60-percent recycled and is available in a minimum order of two cases. Free samples and catalogs are available by calling 617-227-9901.

- Earthcare Paper in Wisconsin is a rapidly growing company that offers everything from recycled paper to gift wrap. They have a 20-pound bond for duplication and high-speed copiers that has a 50 to 70-percent recycled content. They offer a free catalog and samples, and also have volume discounts. Reach them at 608-256-5522.

The laser printer is a powerful tool, and combined with the right supply of paper, the output can be very impressive.

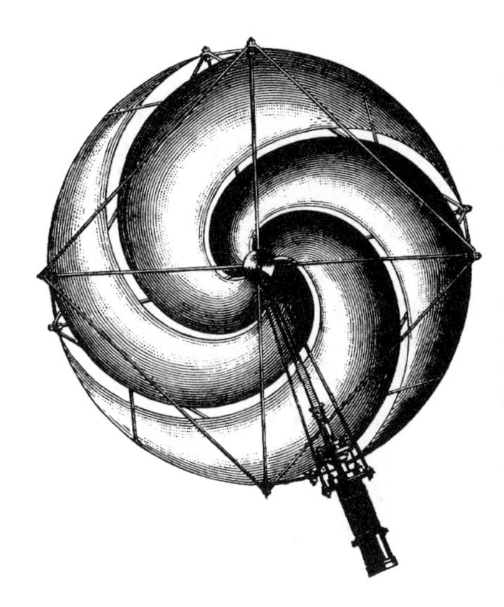

Laser Checks

Make Your Own and Save Money

By Les Cseh of CHEQsys

From *The Flash*, Volum 4, Issue 4

In these tough economic times, we all need to cut costs and improve productivity. Firms of all types and sizes have found that they can save time and money by printing their own checks. New products now allow an entire check to be printed at your place of business, including the logo, signatures, and the MICR (Magnetic Ink Character Recognition) encoding. Let's look at how you can do it yourself, what you need, and what you need to know before investing in these products.

MICR (pronounced either "my-ker" or "micker") encoding is the row of unusual characters at the bottom of checks. These characters encode the account number, bank transit, and check number. When properly prepared, checks with MICR encoding can be processed automatically by the bank's equipment, costing the bank very little per check. When a check cannot be processed automatically, it costs the bank $6.00 to process it manually.

Although not difficult to use, the technology is poorly understood. The result is considerable misinformation, and a growing problem for the banks. Therefore, it is important that anyone planning to use these products realize that there is some education required, there are regulations that need to be followed, and that there is a need for ongoing quality control.

Why Print Your Own Checks?

Why bother printing your own MICR characters? The best starting point is to picture your current check production process. If you or your staff are hand-writing checks, then you know how time-consuming and non-productive this task is.

Many firms use a computer to generate and track checks, printing them on a computer printer. Every time you have checks to print, there may be several steps involved: the current forms must be removed from the printer; check forms retrieved from storage; checks loaded into the printer; alignment tested and adjusted as required; checks printed; checks removed from the printer and stored; normal forms loaded; checks burst/separated; checks signed.

Now imagine that you have several accounts to manage. Depending on the type of business (e.g. payroll service, property management), some companies have 30, 50, and even 100 accounts. Imagine the office space tied up just storing check forms!

The more accounts and/or the higher your check volume, the worse it gets. In addition, you have to keep track of how many checks you have in stock, as the lead time for check forms can be weeks (bad news if you've run out early!). Add to this the fact that check forms have been historically very expensive, a trend which is changing.

A laser printer system for printing checks can overcome most (if not all) of these problems. To start with, check forms designed for laser printers tend to be much less expensive than those for impact printers (dot matrix and daisywheel). A single check form can be used for all accounts, by delegating the printing of company information (logo, name, address), bank information (name, address, routing), and MICR encoding to the laser printer. These systems offer additional benefits as well. One popular feature allows the laser printer to sign checks for you. If this concept makes you nervous, bear in mind that the software will allow you to specify that the automatic signature(s) should only appear on checks for less than some amount you determine. There is also the option of the security cartridge, mentioned below.

Another accounting advantage is that the laser printer can MICR-encode the check number produced by your accounting software so that the check numbers will appear on your bank statement, simplifying bank reconciliation. Also, you don't have to throw out expensive unused checks whenever you change banks or bank accounts, or move. No more four-to-six-week lead times for your checks to be printed.

Printing Checks With Laser Printers

MICR-capable laser printing systems have the following in common:

Laser Printer. Not every laser printer can be used for MICR printing. MICR toner from a reputable source must be available, and a MICR font should have been specifically designed for the printer. In addition, its paper handling needs to be very accurate, especially that it not skew or tilt the paper. "Curl" is an enemy of the bank's check-processing equipment—therefore, the fewer curves in the printer's paper path, the better.

MICR font. The font must not only be visually accurate, but magnetically accurate as well. This means that the font must be designed for the specific combination of laser printer engine and configuration (for example, print smoothing features like H-P's RET), toner, and paper.

MICR toner. Normal toner cannot be used; it does not have the proper magnetic characteristics. Printers using toner cartridges (such as the H-P LaserJet and Apple LaserWriter) allow you the option of switching between MICR toner for checks and normal toner for other applications.

Security Cartridge. You can use a cartridge (like a font cartridge—not a toner cartirdge) containing the MICR font, logos, and signatures. When the cartridge is inserted into the printer, checks can be printed; when it is removed, the printer can still be used for office applications, but not for checks. Some vendors store this sensitive information on the computer, rather than on a cartridge. This is fine for some, but it may be too much of a security compromise and inconvenience for others.

Check Stock. The paper must be carefully selected to match the specifications of the bank and of the laser printer. The specifications cover the materials, inks, colors, reflective and structural criteria, perforations, and more. The design should be very general, to avoid obsolescence. In addition, the design should incorporate the use of preprinted security features which will deter all but the most serious counterfeiter. Although you could use a completely blank check stock, we do not recommend it: it is risky, and the additional toner required will probably outweigh any savings.

Quality Control. It is vital that anyone involved in producing checks have an ongoing quality-control program in place. At the very least, a thorough visual inspection should be carried out at the start of the check run, and preferably during and at the end of the run as well. The simplest part of the inspection is to ensure that the print quality is excellent. MICR alignment and position are easily inspected using a special MICR "template."

Thorough visual inspection should catch 95 percent of all problems that could occur on a good system. However, it will not catch problems with the magnetic quality of the printing. Be sure that the vendor offers magnetic check testers, and/or a check testing service.

Software. The best scenario is for your accounting/financial application to directly support the specific laser check-printing system you intend to use. However, since few accounting systems fall into this category, third-party check-printing software is designed to pick up check output from your accounting software and rework it for laser/MICR printing. MICR check software is available to work with most micro- (DOS, Windows, Macintosh, etc.), mini-, and mainframe computer systems.

Vendor selection for your check printing system is critical. The components must be carefully matched, and the company behind the products must understand the technology, the issues, and the standards. Has the vendor invested in a MICR analysis system? Do they offer all of the required pieces, and if not, how can they assure that their piece will work properly with the others, now and in the future? Your vendor should have worked with the financial institutions in the development and testing of their products.

Laser printers have proved their suitability to the task of check writing. Many organizations have freed up considerable time and money that used to be wasted in the production of checks. With time, technology keeps improving and giving us more time.

Les Cseh is managing director for CHEQsys, a firm specializing in complete check-writing and check-printing solutions. CHEQsys may be reached by telephone at 416-475-4121 and by e-mail on CompuServe at 76424,2075.

Making Rubber Stamps

From Laser Copy to Ink Pad in Three Easy Steps

By Holly Blumenthal

From *The Flash,* Voume 4, Issue 4

The latest addition to our grab bag of laser fun is a Polly Stamp machine from Grantham's of East Grand Forks, Minnesota.

The Grantham Polly Stamp system takes the text and graphics from a laser printer and creates colorful self-inking or block-style rubber stamps. It's a blast, and you could even develop an at-home business around it: desktop-published rubber stamps! In a nutshell, you design an image on your computer screen using a word processor, drawing program, or what have you. The final image can have several different stamps arranged on the 4.5-by-9-inch area available. Print a negative of the image on two separate transparencies, along with registration marks. Align the images via the registration marks, add goop (Figure 1), and expose it to intense ultraviolet light (in a sealed, light-tight box). Presto: The goop, a light-sensitive resin, hardens in the areas exposed to the ultraviolet light.

After cleaning up the flexible rubber stamps, trim off any excess rubber, and attach the stamp to the plate of a self-inking stamper or wood block. Ta-daaa! Your computer graphics are now ready to apply to any flat surface including envelopes, invoices, stationery, wallboard, wrapping paper, your forehead...you name it!

It is possible to make good stamps with the first try, but as with any craft, understanding and experience will be certain

FIGURE 1
Adding the goop for rubber stamps

allies in working with the Grantham Polly Stamp system. The stamps cannot mimic the detail that is possible with a laser printer. The rubber material has limitations in terms of point size and line width that will produce a good stamped image. There is also a skill to determining the amount of exposure that will create a base that is not too thin and not too thick. A thick base does not leave enough relief between the image and the background. If the base is too thin, it will not support the fine details. Grantham's manual provides helpful basic guidelines for successful first attempts; and making a few stamps that don't work well is surely a good way to learn about the possibilities and test the boundaries.

We recently used the Grantham Polly Stamp to create a variety of self-inking stamps for BlackLightning. Previously we had sent out for rubber stamps of our address, check endorsements, and special things like First Class or Book Rate mailing stamps.

The only frustration we ran across was with the self-inking stampers. We were stamping hundreds of pieces of literature, and it seemed that the stamper pads frequently needed reinking, which was a bit of a bother. This may not be as much of an issue for those who are only using the stamp for a few things a day, but the next time around, we will probably go for the simple, old-fashioned, wood-block stamp and a stamp pad. For our needs, it seems that simplicity works best. An advantage to the old-fashioned block stamp is you can easily change colors by simply getting another color ink pad.

If you only need a few stamps, the Polly Stamp (like any significant piece of equipment that costs $595) is not an economical buy. But if you need a lot of stamps, want to start a new service, or just want to have fun, then the Polly Stamp machine is a great accessory to your laser printer! Grantham has two packages available. Both have the machine and all the materials needed to get started. The first is a new Polly Stamp exposure unit with starter package, for $965. The second package has the same equipment and supplies, except that the Polly Stamp exposure unit is refurbished, not new, and the package price is only $595. (Both prices quoted include shipping.)

There are certainly all kinds of fun possibilities with this stamp-making system, and the cost would be a minor investment for a new business venture or an adjunct to an existing business. The machine and process take very little room; everything can be done at a small kitchen table. The stamp making could be nice diversification for a wide range of established businesses including desktop publishers, graphic artists, specialty shops, custom printshops, general stores, and mail order businesses. The venture could also include products made with stamps. Stationery and wrapping paper decorated with stamps have a particularly charming style. Stamps might be designed and made inhouse for standard stock, as well as offering customized stamp making. It seems that when you are looking to have a custom stamp made, you usually are limited to text only. If you put your computer and laser printer together with a tool like a stamp maker, your imagination is the only limit!

Further Information

Grantham's Polly Stamp
418 Central Avenue NE
East Grand Forks, MN 56721
218-773-0331

Getting Transferred

Using Transfer Toner for Color Applications

By Walter Vose Jeffries of BlackLightning
From *The Flash,* Volume 1, Issue 2

In July of 1989, BlackLightning introduced a radically new type of toner to the laser printer market: Iron-On Transfer Toner. We began shipping blue toner in the beginning of August, and the response has been overwhelming! We are now shipping twelve colors with an improved formulation, and continue to make advances. We have received a tremendous amount of positive feedback on the transfer toner. This article will give you a little background and pass along some discoveries that people shared with us in the four months after Transfer Toner was introduced.

The sublimation process has been used for years in the clothing industry to transfer images to fabric, and in the trophy business to create inexpensive, quality plaques. A number of vendors offer systems based on copier technology, that allow people to photocopy an image and produce an image that can easily be transferred to fabrics and coated metal surfaces with a heat press. The problems with these systems are that they typically have significant amounts of background spotting due to the copier technology; they are expensive; and more importantly, they lack the powerful typesetting capabilities of a computer and laser printer system.

Multicolor, multipass printing is easy to do with transfer cartridges, as all of the toner components are completely contained within the cartridge, and cartridges can be easily and rapidly changed. Using the blue (cyan), magenta, yellow, and black cartridges, you can even do four-color process work to produce color photographs!

The technology involves embedding special, expensive sublimation dyes used in the fabric industry into the toner during the manufacturing process. The toner acts as a carrying agent for the dye, and for the most part is not transferred to the final product surface. The color comes from the dye, not the toner. The result is an image that looks black on the paper (plain paper is used), and shows its colors when transferred. When the transfer is heated to 400°F the dye *sublimates* (changes from a solid to a gaseous form) and moves to the new surface. The dye bonds to polyester molecules in the new surface, producing the color. This process takes between 15 and 45 seconds, and produces color images fully washable in hot or cold water, even with bleach. It can be used with any surface containing polyester, including fabrics, coated metals, painted surfaces, PC circuit boards, and coated ceramics.

Equipment

A laser printer with a Canon-based printer engine is the basic hardware needed to produce transfers. This includes laser printers from manufacturers such as Apple, Hewlett-Packard, QMS, Canon, and many others. You will also need a computer, such as the Apple Macintosh or an IBM compatible, to control the laser printer. To transfer your images you will need a household iron or better yet, a T-shirt press or similar equipment. Many people are finding scanners to be useful in getting existing artwork into the computer quickly.

Software

A wide variety of software on the market is useful in helping you design your transfers. At BlackLightning, we use Adobe Illustrator and Adobe Photoshop extensively. These packages' powerful drawing and color separation capabilities are ideal. Many customers find Dr. Halo, PC Paint, Corel Draw, and

PhotoStyler to be very useful on IBM-compatible computers. Most paint or drawing programs will do, and a word processor is all you need for simple typesetting. The key feature is being able to flip your image horizontally so that it will print in mirror image. On a Macintosh, simply select "Flip Horizontal" in the Page Setup Options dialog box (Figure 1).

FIGURE 1
Page Setup Options dialog box, Flip Horizontal

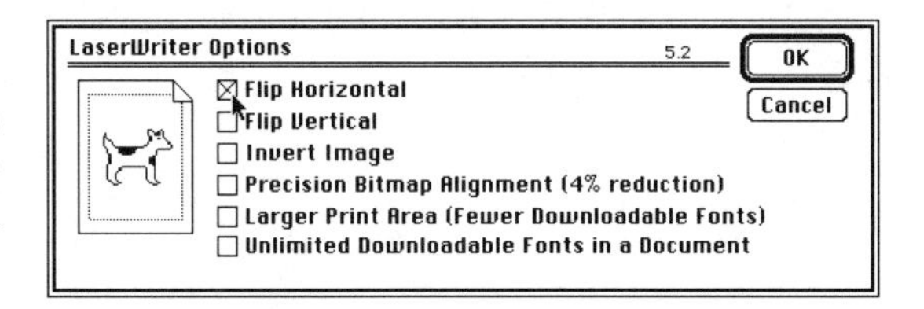

Square Screens

Jim Reppond from Kobe, Japan, wrote us with a tip on getting the best photographic images from your laser printer. In Photoshop's Page Setup dialog box he chooses "Use Accurate Screens" and "Square" shaped screens. This same trick can be used with PhotoStyler on the PC, and with many other programs that let you adjust the lines per inch and other screen parameters. Beware that "Accurate Screens" is a feature of Adobe PostScript Level II, and may not be available in printers using other page description languages.

Designing your graphics

The most important issue when designing your graphics is to consider the surface you will be transferring to. The resolution of your image is limited by the surface of the destination material. If you are transferring to a metal or ceramic surface, or to a very tightly woven fabric, small details and type will show up well. If you are transferring to a fabric with an open weave, you will need to have larger graphics to produce clear results. Don't transfer 9-point type onto a regular T-shirt, or it may be unreadable. When transferring to a smooth surface, expect the image to be very similar to what you get from the laser printer; when transferring to a coarse surface, you need bigger, bolder lines and images.

When transferring to fabrics, the range of screens or halftones that can be achieved on a laser printer may not show on the final transfer. You may not be able to differentiate between 5-percent and 15-percent screens, whereas you might be able to on a standard laser-printed graphic.

Whether you start with a scan, create a drawing directly on the computer, or typeset a message, be sure to reverse the image before printing.

Printing

Before you create your transfer, proof the image with a non-transfer cartridge to avoid wasting transfer toner. You may also wish to iron the transfer, done with transfer toner, onto a scrap of material before transferring to the final surface. We strongly recommend that you experiment before attempting your final product. You can use regular copy paper for your transfers. No expensive, special papers are needed.

When you are ready to print, load in the transfer cartridge of your choice and the fuser wand that came with it, and print a couple of copies of your image. The resulting image will appear gray or black on the paper. The transfer will show its colors when you heat it. Change back to your regular cartridge and wand after you have printed. Because of the cartridge system used by Canon based laser printers, you can easily swap cartridges and not have to worry about residual toner being left in the machine and needing to be used up.

If you are printing images with a lot of coverage, you may need to change your fuser wand felts more frequently. Background streaking on the printed output indicates it may be time to change the felt.

Multicolor Transfers

Printing single-color transfers is all well and good, but it is the capability to print multicolor images directly from their laser printer that has many people excited. This can be done simply by passing the sheet of paper through the laser printer several times, using a different color transfer cartridge on each pass.

Not only can you get several colors on a transfer, but you can also mix the colors to achieve entirely new colors, as Frederick Ross, of Terminal Solutions in El Paso, Texas, discovered.

The biggest concern in multicolor/multipass printing is registration. We have found the Canon printer engine to typically have a registration of better than one thirty-second of an inch, using autofeed from the paper tray when the machine is well maintained. Manual feed tends to produce slightly less accurate results.

The first step is to prepare your image for color separation. Some programs, such as Adobe Photoshop, have built-in color separation capabilities. With other programs, you will need to prepare separate files for each color. The easiest way to do this is to create your drawing and save one copy with the extension ".All" on the name. Then selectively delete parts that should not show up for each additional color, and save them with the extension ".Blue," ".Red," etc. If you had a picture of a purple cow, green fields, a blue sky, and a red barn, you would need to save the files illustrated in Figure 2.

FIGURE 2
Files for color separation

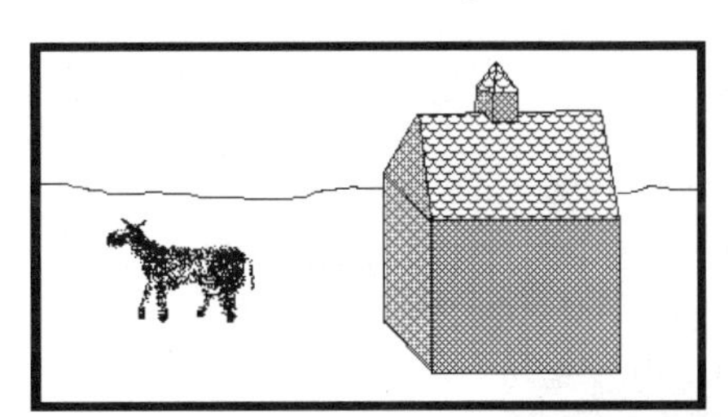

Pastoral.red
(the cow & the barn)

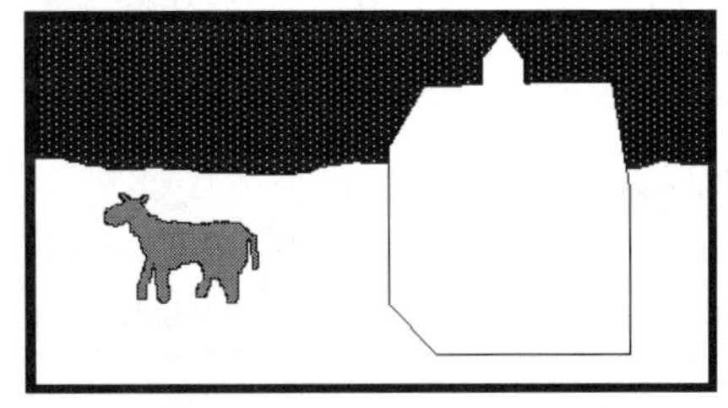

Pastoral.blue
(the cow and the sky)

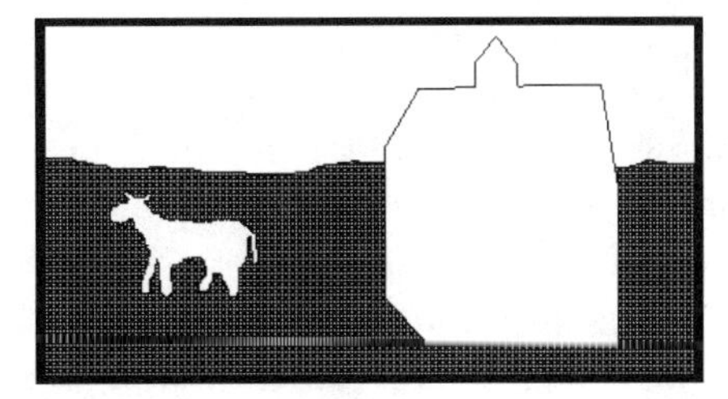

Pastoral.green
(the fields)

Printing each of the files in turn, changing the transfer toner cartridge to the appropriate color, and reinserting your paper in the automatic paper feed tray for each pass will give a you five-color transfer. The fourth color is purple (red and blue) and the fifth is black (red, green, and blue.)

Experiment with mixing colors and registration. Expect different sections of the page to register better than others. Try placing a black outline at the border where two colors meet, to hide any misregistration. Mixing red and green will produce brown, while red and yellow produce orange.

Registration

When creating multicolor layouts for use with color toners or transfer toner, good registration is often critical. There are a number of techniques for improving registration. Outlining between colors and overlapping on both sides a little helps define a design and mask slight registration discrepancies. When loading the paper tray for each successive pass, fan the stack of papers and place it in the tray. Tap the tray until the pages rest evenly against the front edge. Be careful not to overheat the machine; that will cause the registration area to change. Keeping your machine's rollers and pickup claws clean will help keep the paper on its path. Prepressing the paper to remove moisture, by passing it through the printer while printing a blank page once, is also good.

Transferring

We'll cover transferring to fabrics here. Other surfaces are similar. As mentioned before, there must be polyester in the destination surface for the full brilliance of the colors to be revealed, and to assure washability. This eliminates pure-cotton fabrics (we are working on a solution to this). 50/50 cotton/polyester blends work well, although 100-percent polyester is ideal. Since no one likes to wear 100-percent polyester, the trick is to use 50/50 and prepare it with PrepSpray, a polyester solution, that you can spray onto the front or back of the shirt and thus increase the image brightness significantly. We have found that Screen

Stars' Best T-shirts are especially good. These 50/50 shirts have a tight weave that takes colors better than common undershirts.

If you are using a hand-held steam iron, be aware that the steam holes will be cooler than the iron, and you must lift and move the iron during the transfer process to avoid having the steam holes show in the image.

Place half a dozen sheets of paper inside the shirt to prevent the transfer from showing through on the opposite side. Smooth out your fabric and iron it flat to remove any wrinkles in the area you are transferring to. Place your transfer face down on the fabric, and place another plain piece of paper over it to mask it and prevent the fabric from overheating. Customers have reported that a brown paper bag works well. We used blank newsprint. Press the image firmly for 15 to 45 seconds and remove the transfer to reveal your finished product!

Examine the transferred image. Is there spotting in the background? If so, turn the density setting on your laser printer down, or try transferring for a shorter time. If the image is not dark enough, try turning the density setting up, or transfer for a longer time. If you get a haze in the background then you may want to try prepressing: Press the transfer onto a scrap of fabric for about seven seconds at 400°F. This will remove any haze, but leave the main image with plenty of dye to transfer the next time. You'll find that many transfers can be reused a second time.

Washing

Now that you have produced your beautiful work of art, you will may be wondering how to care for it properly to keep it from fading. Our field tests show transfer toner to be extremely resistant to fading on fabrics when properly applied, even after years of hot water and heavy bleach washings.

Applications

When we first introduced the transfer toner, we were primarily thinking of T-shirt applications. Finally the time has come when you can take your gorgeous computer graphics and transfer them to many different materials. Since August, we have received literally thousands of responses from excited customers. People are using the transfer toner in ways we never imagined! While many of you are making up wild and crazy T-shirts (and thanks to all of you who have sent in examples of your artwork already!), others are making mugs, plates, personalized baseball caps, the labeling on computer circuit boards, front panel displays for electronics hardware, trophies, plaques, leather appliqués, designs for embroidery, craft patterns, stenciling on sheetrock, and many more creative applications. Several people have opened new businesses or expanded existing businesses utilizing transfer toners, and educators across the country are using the transfers in fundraisers. In fact, one gentleman bought his second and third transfer cartridges with the money he made selling the transfers from the first one! The possibilities are limited only by your imagination.

Materials and Product Ideas

We have transferred to a wide variety of materials and products, including the following.

Apparel. 50/50 cotton/polyester T-shirts, sweatshirts, sweat pants, tank tops, and other apparel. Whites and light colors look best. Double knits produce even better transfers.

100-percent Polyester Double-Knit Fabric. Excellent for making patches, flags, and banners. One user had a 12-by-12-foot photo on a banner!

Baseball Caps. Ones with 100-percent polyester front panels take transfers well, giving rich colors. Light-colored, white, and neon caps look fantastic.

Tote Bags. Use PrepSpray on canvas tote bags. Encourage people to use tote bags rather than paper bags when shopping, to help save trees.

PVC. Pipes and flat panels for making signs, tabletops, and coasters.

Open-Pore Anodized Metals. Works, but the dye is just sitting in the pores and is not bonded.

Specially-Treated Metals. Brass, steel, and aluminum are available with a very thin surface coat of polyester. These make excellent signs, desk name plaques, trophies, awards, specialty business cards, classy luggage tags, Christmas ornaments, and more. Coated metals are available from a variety of manufacturers and come in a wide range of quality. Some will even resist ultraviolet (sunlight) fading, although none we know of will last for years.

Ceramics. When mugs or plates are specially coated with polyester, they produce brilliant results. Coated mugs require special presses and are available from a number of suppliers—and can be found in advertising in *Trophy* and *Award* trade magazines.

Leather. If it is untreated (no polyester), then the images are not as brilliant and could wash out, but some interesting effects can be achieved.

Velcro. Bands were tested successfully for a special project by one customer.

Plaques. These are available in a wide variety of shapes and sizes, and can be nicely mounted on wooden backs. Premium metals are the best and should be used where quality counts.

Business Cards. Novel business cards from metal! Great gift and award idea!

Brass Christmas-Tree Ornaments. Interesting shapes like teardrops make great Christmas ornaments and gifts. Personalize them to keep the memories alive.

Key Chains and Luggage Tags. Metal and PVC can be used to make classy personalized tags. Especially beautiful in brass.

Dog Tags. "My name is Spot. Please tie me up and call 825-555-1212."

Medallions. Round metal discs or "coins" make great medallions. Ideal for promotions: e.g. Collect five and get a free pizza!

Signs. Metal, vinyl, PVC, or 100-percent polyester fabric. Bigger signs can be done with multiple pages pieced together.

Refrigerator Magnets. Metal, vinyl, or PVC can be transferred to, and then attached to magnetic backing for fridge magnets and removable car signs.

Placemats. Vinyl placemats are great for kids. They're washable. Harder vinyls are better. Very soft vinyls may bleed the colors over time.

Bumper Stickers. Special stickers will accept transfer dyes. We've seen them with iridescent backgrounds and in flourescent colors. We hope to line up a source of materials for this in the fall.

Patches. Transfer to 100-percent polyester; sew to iron-on glue backing, or stitch directly to clothing to make brilliant custom patches.

Coupons. One customer uses a custom mixed color of transfer toner to make unforgeable gift certificates. This rates as a most unique use of transfer toner.

Labeling Circuit Boards. Most electronic circuit boards will accept transfers for labeling where components go and who manufactured the board. Transfer toner is not conductive, so it cannot be used to lay down the traces. People have had success using the graphic toner to mask against the acid baths, and then using transfer toner to label the final circuit board.

Equipment Front Panels. A number of engineers are using transfers to make up the front panels for mockups and limited runs of equipment displays. We have seen several creative and beautiful examples of everything from mockup panels to small-scale production runs showing all the lettering for the dials as well as the company logo, artfully done.

Wallboard. Sheetrock can be high-tech "stenciled" with transfers.

Tapestries and Banners. Several people have used transfer toner to make large banners and tapestries measuring four-by-30 feet, 10-by-10 feet, and larger. The computer enlarges the image and prints it on many pages. Then just transfer in the proper order.

Shower Curtains. Make a wild shower curtain of double-knit polyester with dozens of colorful designs. The double-knit material is thick enough to stop the shower spray, yet it feels soft and looks much better than the tacky clinging plastic shower curtains. We sewed a dog chain into the bottom to keep it from billowing into the shower. Easy to clean, too. Just throw it in the wash!

Window Curtains. Several *Flash* readers wrote about making window curtains for their home, one using kids' drawings.

Handkerchiefs. Easy to make; just transfer and trim. A great gift idea.

Kids' Clothing. Make clothes, and then transfer a design that can even go across the seams for better-than-professional-looking workmanship.

Painted surfaces. These will often take transfers, too.

This is by no means an exhaustive list. Try new things. The basic characteristics are that it must not melt, burn, or warp at the required 300°F to 400°F. A high polyester content is a definite plus and will produce brilliant, long-lasting images. Buy small quantities of some materials and try them yourself. Share your results with others. There's a world of opportunity.

Other Sources of Information

Get a subscription to *The Flash* (Riddle Pond Road, West Topsham, VT 05086), *Impressions* magazine (Gralla Publications, PO Box 801470, Dallas, TX 75380-9945), and to *The Engraver's Journal* (PO Box 318, Brighton, MI 48116).

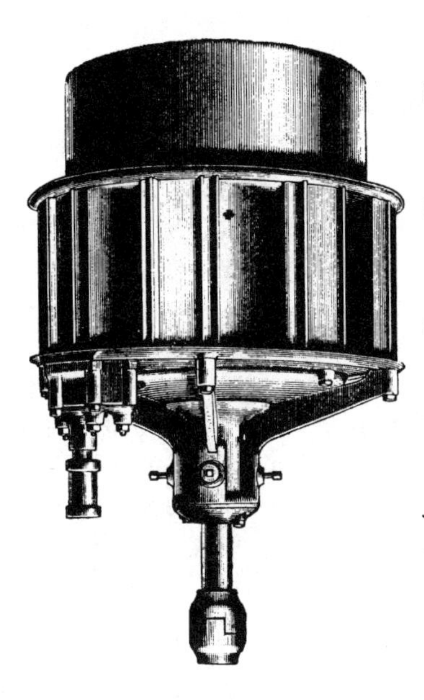

Tricks of the Transfer Trade

Tips and Techniques for Using Transfer Toner

By Walter Vose Jeffries of BlackLightning
From *The Flash*, Volume 3, Issue 2, and Volume 4 , Issue 1

Here are a bunch of methods and ideas we've discovered for working with transfer toner that really defy classification.

How can I transfer to 100-percent cotton? Use PrepSpray to add a thin layer of polyester to the image surface of the fabric. This can also be done to 50/50 shirts to improve the brightness and density of the resulting transfers. The new Power Pak Prep-Sprayer applies an even coat to fabrics, metals, ceramics, and other surfaces. (Note the PowerPak does not use CFCs.)

What is color mixing, and how do I go about it? Color mixing refers to using two or more base colors to produce a color that is not available in a single cartridge. For example, red and yellow transferred to the same area will produce orange. Some basic combinations are: Purple = 50% Red + 50% Blue; Orange = 20% Red + 80% Yellow; Brown = 50% Red + 50% Green. By adding black you can make the colors darker. Different materials will show the colors differently, depending on the material's polyester content and color. For example, yellow on a blue shirt will look greenish. Make samples for yourself, mixing the colors that you have in different combinations. Use these to know what you'll get when you go to do a project.

Darker than Dark. Print the same color twice to get extra rich dark colors. Blue + Blue = Navy Blue.

Should I remove the plastic film from coated metal I purchased for transferring? The clear sheet protects the finished surface of the metal from scratches. You can remove it before or after you transfer. If you are doing fine detail work then you may need to remove it before transferring, as it can blur the image a little. I normally transfer right through it and then let the piece cool completely before removing the plastic sheet. This eliminates the need to use TransClean to remove the toner from the surface of the metal. If you remove the plastic sheet before transferring, let the piece cool completely, peel the paper off, clean the metal with a dab of TransClean on a wad of toilet paper, and then buff it with clean toilet paper.

Making Your Own Flat-Iron Heat Press. Transfer Toner is not just for professionals and commercial use. A lot of small businesses, departments, and individuals are using it to make custom shirts, hats and signs. Heat presses are great if you have the money and are doing a lot of shirts, but a fair number of people are using household steam irons to press their transfers. The problem with a steam iron is that it has vents in the bottom which create cold spots, resulting in light areas on the transfer. This can be reduced by carefully shifting the iron. A better solution is an older flat iron without steam vents. These have become very hard to find, and most are old and worn out. However, you can make your own. All you need is a steam iron and a sheet of thin metal (Figure 1). I used 1/32-inch-thick uncoated aluminum. Place the iron flat on the metal and trace the foot with a nail. Next make another tracing about one half-inch out from the first, and cut along this last line. Then cut slits from the outer to inner tracing, so you can easily fold the metal edges up

FIGURE 1
Making a flat-iron heat press

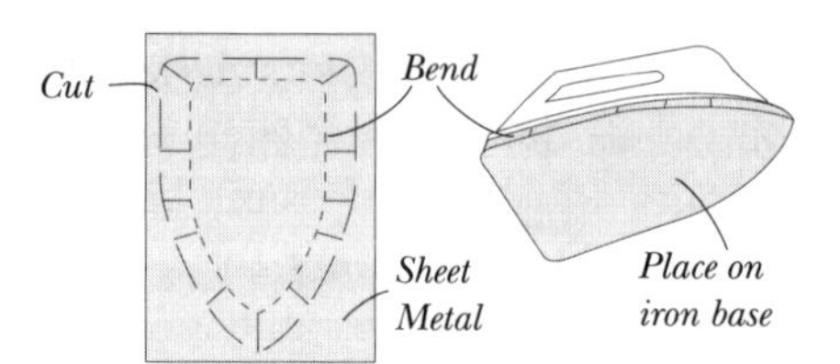

along the sides of the iron's foot. Hammer them gently to get a secure fit. By bridging the vent holes this way, you can get an even heat across the whole iron.

Be Cool! Let your transfers cool completely to room temperature before lifting. You'll get less toner on the material, and the paper won't rip (extra important on fabrics and some metals). Mugs, however, should be stripped at their hottest. Then dunk them to cool the surface and prevent the dye from migrating and blurring. If the paper sticks, the mug was too cool. Use TransClean and a wooden popsicle stick to remove the adhered paper you can't pull off *after* cooling the mug. (If you don't strip the paper from the mug prior to dunking, it will adhere tightly to the mug surface.) If your mugs are too hot (>475°F), then scoop up a little water and swish it inside the mug for a moment before dunking the whole mug. This will help prevent cracking. After the mug or metal has cooled completely, then clean off the residual toner with toilet paper and TransClean.

Premium is Better. We find that the premium mugs have a much smoother, more even surface than the promotional mugs. This translates to the ability to transfer from the top to bottom edges on premium mugs.

Custom Heat-Press Pads. When applying transfers to already decorated shirts or doing the cuffs on sleeves, we needed custom base pads for the heat press. Silicon rubber (it's expensive) works, but you can also get good results using the dense black foam used to pack many products. Test a small piece by pressing it between two sheets of paper, to make sure it can withstand the high temperatures needed for transferring. Cut it to size and slip it inside the shirt cuff or body. This gives you a raised, smooth surface to transfer to, so you can apply heat to just the area of interest.

Cleaning. If you are using the premium UltraCoat metals and pressing through the plastic film, be sure to clean off the pink lettering on the film first, or it may transfer to the metal. Use acetone, Fedron, or BlackLightning Cleaning Fluid on a piece of tissue.

Fuser Lamp Wattage. Some customers have had problems with their transfer toner because the fuser lamp in their printer is too hot (Figure 2). The fuser lamp is replaced from time to time as part of routine maintenance. Many varieties will work with regular toner, but some are too hot for transfer toner. The fuser temperature is controlled by a temperature sensor that turns the fuser lamp on and off to heat the roller. If the lamp power is too high, it will tend to overshoot the desired temperature. This overmelts the toner, causing excessive buildup on the fuser wand, and streaking on the page. If this is a persistent problem, the only solution is to replace the lamp with one of the proper power.

Keeping Clean. Transfer toner is different from regular print toner. If much toner builds up on the felt cleaning pad, it can cause backgrounding, streaking, and offsetting. Printing some blank pages between jobs will help keep the rollers clean. Each transfer cartridge comes with extra cleaning felts, and you can purchase more if you need them. These felts are thicker than the felts shipped with other cartridges, so be sure to use only transfer toner felts when printing with transfer toner. One trick for minimizing the buildup of toner is to position large images diagonally on the page. Avoid the long orientation, as it overuses one spot and doesn't allow the printer to clean itself as well as with horizontal or diagonal orientation. Even a slight angle off of vertical can make a big difference in preventing streaking.

Mug Preparation. When we have many mugs to produce, it is important to set them up as quickly as possible. We use two boxes about one inch high, spaced about 10 inches apart on a table to hold the mug while we tape a transfer to it. We place the mug with the handle resting on one box and tape one side, then roll the mug handle to the other box, and the opposite side is ready for taping.

FIGURE 2
Bad pages

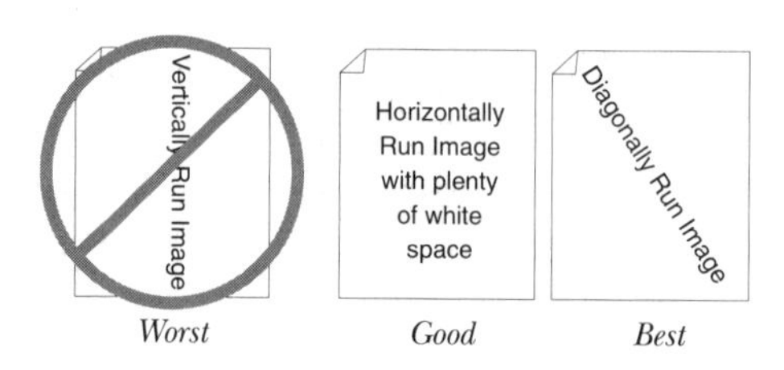

Proofing. Use regular black and color cartridges to proof work in multiple colors. You can now get seven colors of cartridges from BlackLightning that will print color on paper.

Prepressing To Eliminate Backgrounding. Paul Brehm of Ohio suggests a way to eliminate backgrounding and get maximum color saturation. Press the transfer at 400°F for 10 seconds onto a scrap of material (we use polyester). Let the transfer cool to room temperature before removing it from the fabric. Then do the real transfer for the maximum time the material will take. The first press uses up the dye that was spread by the fuser rollers, letting you transfer a lot longer the second time without getting backgrounding. This technique is useful with all materials, mugs, T-shirts, metal, etc.

Keeping On Center. Howard Eglowstein of *Byte* magazine has a good way to mark the center of a piece of fabric. Fold it twice, and pinch the inner corner. This will leave a crease. Center that crease on your heat press, and use it to center your transfer on the material.

Transparent Transfers. David Bueschel in Wisconsin has a way to create color transparencies for presentations, pseudo stained-glass windows, and transparent window signs by transferring onto transparent polyester film such as *Avery Transparencies.* Clean the film with TransClean. For a dark black, transfer the other colors first, and clean the transparency. Then, on a separate piece of paper, print the black pass using graphic toner. Press this graphic black image onto the transparency for 40 seconds to produce glossy signs. Or, do the black by simply printing with graphic toner directly to the transparency after all the other colors are pressed on and cleaned up.

Matte Finish On Plaques. You can get an impressive matte finish on plaques by using Graphic Toner rather than Transfer Toner. Here you are actually transferring the toner rather than a dye. The trick is getting the toner to release from the paper. Try experimenting with freezer paper and the backing for Avery labels. This same technique may be applicable to making printed circuit board etchant masks.

Full-Color Photo Transfers

Using Transfer Toner for Color Photographs

By Walter Vose Jeffries of BlackLightning

From *The Flash,* Volume 4, Issue 2

We have received a tremendous number of requests for information on doing full color photographs with transfer toner. We've not yet perfected the process to our own satisfaction, but due to the high interest level, we thought we'd share our findings to date.

Our objective is to produce near photographic quality, color accurate transfers. By "photographic quality" we mean that the dots used to create the image should be minimized and the shading should be as close to continuous as possible, avoiding the stepping and overexposed highlights normally associated with computerized photos. By "color accurate" we mean that the colors in the final transfer onto white polyester should be as close as possible to the original photo.

The following is the result of our experimentation over the past year. It is based on using Macintosh software, but you should be able to easily generalize the techniques for use with PC computers and software like PhotoStyler, if that is what you have. Most higher-end PC and Mac photo-retouching software that does color separations has similar tools. Unfortunately we can't help you with programs other than Photoshop, as it is the only one we use. If you get stuck, try consulting your manual's index and table of contents. Try synonyms. Last, call technical support at the company that makes your software, and possibly fax them these instructions.

The downside of doing color separations is that it takes a lot of memory and processing power to do large full-color high-resolution images for T-shirts; even a small image for a mug can use up a lot of memory. For our work we use a Mac IIci with 8 MB RAM, 8-bit video, 105 MB hard disk, Adobe Photoshop 2.0, a 24-bit-color 600-dpi scanner, and a LaserWriter with a Xante Accel-a-Writer upgrade for speed and increased resolution. We maintain about 20 MB of hard disk space free for image manipulations and temporary files. With this setup and practice, we can take a postage stamp or a color wallet-size photo and blow it up to a full-size T-shirt image in about five minutes.

Accurate color matching is still a hurdle we have to cross. The procedure we outline below is designed to give good color matching with prepressing, but is not exact. The prepressing is very important, as that is where you can increase the resolution of your image by dithering it to almost-photographic quality rather than a computerized look with visible dots. The dithering is achieved when you heat the transfer, causing some of the dye to vaporize, migrate, and fill in the space between the computer dots. This softens the graduations between shades. This mixing of the dyes does not happen at the same rate for all colors, so the first image will look too dark and the color will be way off. But the next press should give good color, and excellent contrast and brightness.

Step-by-Step Photographic Transfers

1. The first step is to scan the photo at the highest resolution you can with the available memory. The reason for starting with the highest possible resolution is that once the image is in the computer, you will be better able to edit it, clean it up, and then easily reduce the resolution for the final image. Once you've scanned it, you can't increase the image resolution without getting jagged edges.

2. We overscan a little, to make sure we get all of the image. The next step is to crop the image down to just

the rectangle we will use. By reducing the image size, we save memory and speed up the program.

3. Rarely is the original the exact same size as the final transfer. Once you've cropped the image, resize it, and set the desired resolution for the final transfer in lines per inch. For our purposes here, unlink the file size (resolution or dots per inch—dpi) so you can separately adjust both the size of the image (height and width) and the dpi. Each dot will later be translated to an imaging line, so the dpi here is the equivalent of lines per inch (lpi) in the next step. Hard surfaces need a finer resolution and rough surfaces like fabrics can take a lower resolution, giving more levels of shading.

4. On the Macintosh, the Page Setup Options dialog box on a PostScript laser printer allows you to flip the image left to right so the transfer will press correctly. Photoshop can also set the printer dpi and lpi field information in the Page Setup dialog box. Note that a larger number in the lpi field will give you finer lines, but fewer levels of shading. A lower number gives more shades, but a chunkier, coarser image. We find that with a 300-dpi printer, 30 lpi is good for T-shirts, and 48 lpi is good for mugs and metals. For our 600-dpi printer, we use 90 lpi for both.

5. I like to visually check the image size at this point. In Photoshop, just click the mouse in the file size field in the lower-left corner of the image window to get a size preview.

6. Save at this point, just in case you may want to come back to this point later.

7. The transfer dyes are very strong, and you'll need to adjust the lightness, contrast, and brightness of the image considerably to make it print properly. The Levels tool (see Figure 1) allows you to adjust the brightness and contrast in an image as well as the gamma, or midtones. The Levels tool differs from a contrast/brightness tool in that you can adjust the brightness values of the pixels with middle-gray values

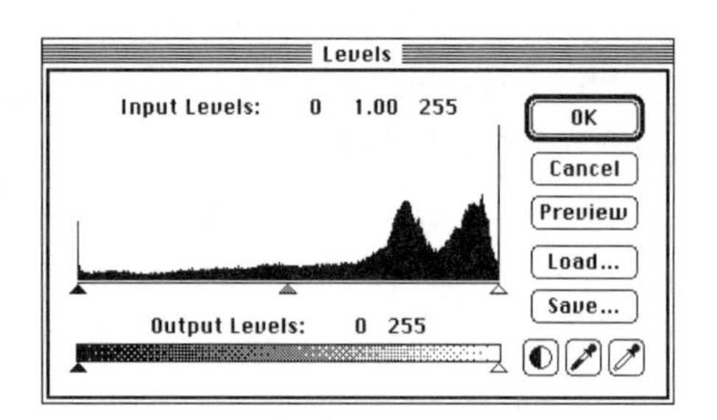

FIGURE 1
The Photoshop Levels control

without dramatically changing the darker or lighter pixels. This setting works best with images that have good contrast. If the photo has poor contrast, try adjusting the Lightness setting two or three times, or try cropping out the background and working just with the foreground image. We sometimes crop out the background and replace it with a graduated or starburst fill. Play with it, and get a feel for how each control adjusts the image.

8. Sharpening can enhance the image and give better definition between areas such as a person's clothing and the background.

9. Photoshop has an impressive array of tools for adjusting color, and for separating colors into their basic components (Figure 2). I feel I have barely scratched the surface of these tools in the many months I've been using this program, and this is the section I expect to keep improving with more experiments. What I have found works best is to do a three-color separation, ignoring the black in the CMYK (Cyan, Magenta, Yellow, Black) color separation. Color theory is beyond the scope of this article. Try the settings mentioned

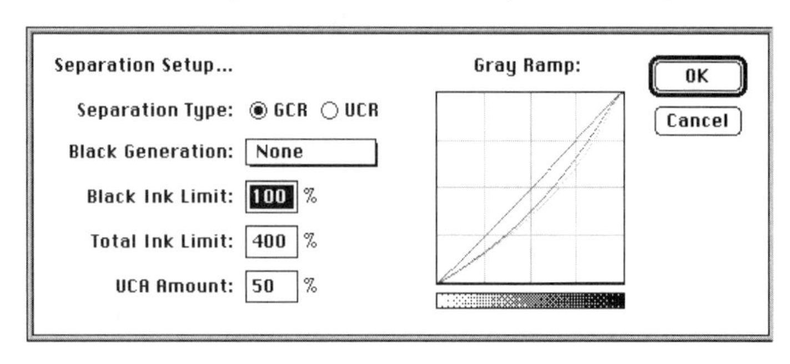

FIGURE 2
The Photoshop Separation Setup dialog box

above and experiment. Gray-scale images, of course, don't need this step or the next.

10. The Hue and Saturation tool (Figure 3) is under "Adjust" in the Image menu. As with the Separation Setup dialog box, I have settings that I have found work well, but not yet perfectly. I use the Lightness setting to lighten up the image. When I have an image that doesn't follow my standard prescription, that's too dark or off color; then it is time to play with the other controls. Suggestion: If you have a section where the color is wrong, make a file with that section repeated several times, then try various adjustments, recording your attempts and results. This is much faster than trying to adjust the whole image at once. On one design, I selected the parts that were not color correct and adjusted those separately from the rest of the image. But usually I apply this tool to the whole image.

FIGURE 3
Photoshop's Hue/Saturation dialog box

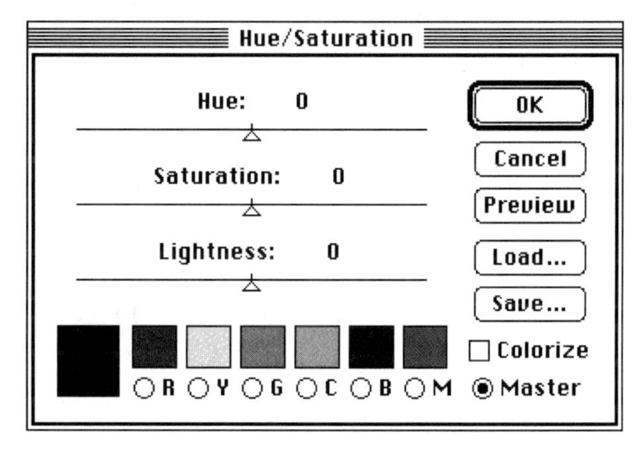

11. Save your work in a new file, using the Save As command.

12. If I'm doing mugs or other small designs, I'll often gang the images, several to a page (Figure 4). Be sure to leave plenty of white space (50 percent or more), as the white space serves to clean the fuser rollers of excess dye. Some positions on the page may have poor registration.

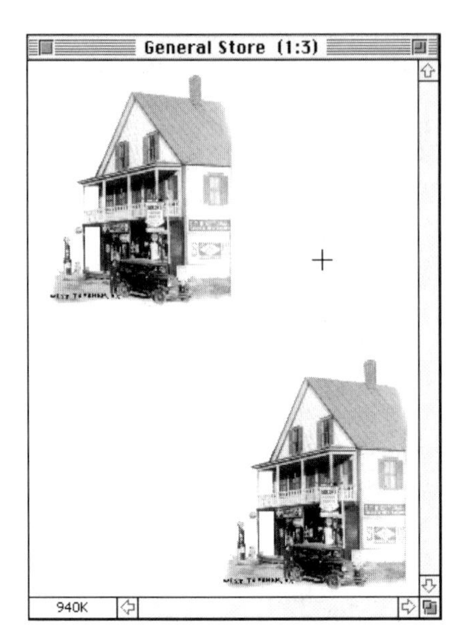

FIGURE 4
Ganging multiple images on a page

13. Photoshop considers each of the color separation levels to be a separate page. There are four pages: Cyan, Magenta, Yellow, and Black (which we are not using). You are printing each color pass for the transfer onto the same piece of paper by passing it back through the laser printer. Registration is critical. If you get poor registration, the colors may be off, and the image will look a little blurry. Get good registration by trying the following tricks.

- Use a CX (EP) or SX (EP-S) laser printer. The newer, cheaper LX (EP-L) personal laser printers have a very contorted paper path, and I have found they do not give very good registration.
- Keep your machine clean and well maintained. Dirty and worn rollers slip easily.
- Learn what parts of the page in your machine tend to have the best registration, and place your image there. I find on my LaserWriter Plus (CX) that the upper-right corner is the absolute best, and the whole right half is generally better than the left half of the page. This is directly related to the next item, which is the biggest cause of misregistration.

- Pre-press the paper prior to printing results in better registration. This removes moisture and shrinks the paper to a stable size. I am looking for a source of reasonably priced, dimensionally stable paper to use, and hope to have a source to mention in a future article.

14. I prepress onto scraps of 100-percent polyester that I purchase by the bolt at a local discount store. It is extremely cheap material and it does the job. For full-color images, I find that a prepress of about 30 seconds dithers the computer dots together, sufficiently mixing the colors and making the image look almost photographic. Even with this extended prepress time, the color saturation of the final pressing is excellent. For single colors or grays, I'll use as little as seven seconds. You can get some very interesting antiquing effects with the black toner by varying the length of the prepress time, because the different dyes used to make black react at different speeds to the heating. Always wait for the paper to cool fully before peeling the transfer off.

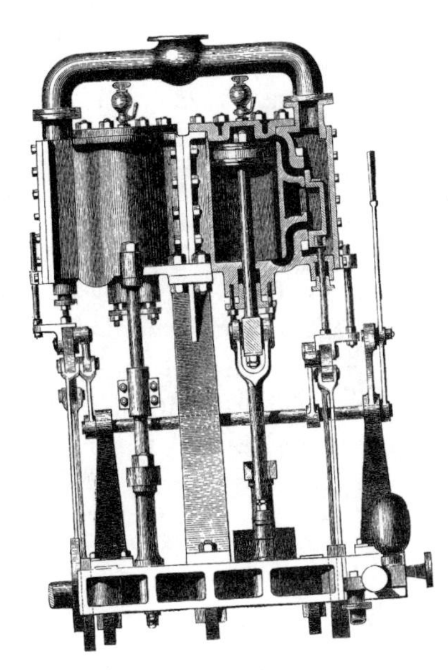

Etching With Toner Resists

Matte Plaques and Printed Circuit Boards

by Walter and John Jeffries
of BlackLightning

We originally developed this technique for creating printed circuit boards back in 1989 at the urging of several customers, including people in Canada and at the University of Fairbanks in Alaska. Obviously, this was a technique with wide appeal. Since then we've used it for a number of other types of projects.

Toner is made primarily of a plastic resin that is resistant to the etching effects of acids commonly used for manufacturing printed circuit boards. By laying a pattern of toner down on top of the copper-clad surface of a printed circuit board, you can create a pattern of copper that will remain after the etching process has been completed by dipping the board in an acid solution. The advantages of using toner directly are twofold: It is more environmentally friendly than the many toxic chemicals used for photoresists; and it is faster, so you save time and money. This same technique can be used for creating plaques and signs. Let's look first at printed circuit boards.

The main hurdle we had to overcome was getting the toner to release from the sheet of paper and adhere to the metal of the circuit board. Using bond paper or transparencies, we were able to get sporadic success, but there were often breaks in the circuit board patterns, making them unacceptable. In the fall of '91 we finally stumbled across a solution that worked perfectly almost every time. Instead of printing on paper, we printed on the peeled-off backing sheets of Avery/Dennison labels. When the labels are removed they reveal a very slick surface that the toner will adhere to lightly, but fully release from when heated and pressed onto a fully cleaned circuit board. Etch it, and voila! one circuit board, ready for populating with components.

The reason this works is that the surface of the label-backing paper is just slick enough so the toner easily releases from it whenever you heat-press it against the printed circuit board, but not so slick that the toner falls off. You could probably make your own release paper by coating it with a very thin layer of silicon oil or some other material. This is something we'll be trying ourselves at some point.

The tricks here are to:

1. Completely clean the copper surface with acetone or something that will remove any oils.
2. Pre-etch the, board or rough it up with a fine steel wool so it is a little rough and will grab the toner better.
3. Print your image in reverse (mirror image) on the label side of a full-page Avery label backing sheet, after having removed the adhesive label paper.
4. Preheat a T-shirt heat press up to 400°F. We have also used a household iron with reasonable results. A dry-mount press might also work.
5. Place the circuit board in the heat press, copper side up, and the page on top of the copper, image side down.
6. Press it for 40 seconds to fully bond the toner to the copper.
7. Let it cool some, but not all the way, say, 30 to 60 seconds, and lift off the backing paper, leaving the toner on the copper.
8. Etch as usual in an acid bath.

Note that the toner will spread a little during the heating process, so your lines should start out a little thinner than you want them to be. Other tips: graphic toner works best; get an Emerald drum, if you can, for the best density and even fills.

A T-shirt heat press works better than a household iron; a dry-mount press might work. Last, in the spring we received samples from two companies that are producing a special paper that may work as well. They are listed below.

This same technique can be used for creating plaques (try doing a negative image so the letters are etched in, or a positive image so the letters are raised), and for making signs for displays on flat PVC signage board. Combine that with Transfer Toner and LaserColor for adding color, and let your creativity run wild.

Sources

Baumwell Graphics
$10 for 10 sheets
450 West 31st Street
New York, NY 10001
212-868-3340
Fax: 212-689-3386

Toner Transfer System
$9.95 for 5 sheets
DynaArt Designs
3535 Stillmeadow Lane
Lancaster, CA 93536
805-943-4746

APPENDIX

Machine Compatibility

Many laser printers use the same basic printer engine, so their supplies and parts can often be interchanged. This chart will help you decipher which toner cartridges will work with your printer. EP cartridges are used in the CX printer engine, EP-S in the SX, EP-L in the LX. These three types can use BlackLightning color and transfer toners.

TABLE 1 **Laser printer compatibility chart**

Manufacturer	Machine	Cartridge
AB Dick	2205	EP
	IP-0800-SMT	EP
Acer Technologies	LP0-75	R4080
	LP-75	R6000
Acorn Computer	L08219	EP
	LX528	EP
	LX3219	EP
	LX3815	R4080
	LX8219	EP
Adex Corp	508	EP
Advanced Tech Intl	800	EP
	870	EP
	880	R4080
	DW-1	R4080
	DW-2	R4080
	GR-2	R4080
	Laserprint 1500	R4080
	Laserprint 1570	R4080
	RP-1	R4080

TABLE 1 **Laser printer compatibility chart** (continued)

Advanced Vision	AVR-LPC3	EP
Aedex Corp	LaserBar 508	EP
	LaserBar-608	R4080
	LaserBar-615	R4080
Alphatext	Alphatext 8	EP
American	Pagewriter	R4080
Apple	Inkjet	Bj10
	LaserWriter	EP
	LaserWriter II	EP-S
	LaserWitr IINT	EP-S
	LaserWitr IINTX	EP-S
	LaserWrtr IISC	EP-S
	LaserWrtr Plus	EP
	Psnal Laser NT	EP-L
	Psnal Laser SC	EP-L
Arkwrite	8A2	EP
AST Research	TurboLaser/PS	R4080
	Turbo Laser Plus	R4080
	Turbo Laser	R4080
	Turbo Laser/EL	R4080
	Turbo Laser/PS	R4080
	Turbo Laser/XL	R4080
ATI	LaserPrint 800	EP
	LaserPrint 870	EP
	Laserprint 1500	R4080
AutoLogic	APS-55/200	EP
AVR	LPC-3	EP
BDS	630/8 LP3X	EP
	630/8 LP	EP
	630/8E	EP
Bedford Computer	QMS 800	EP
Blaser Industries	Blaser LP	EP
Brother	HL-8 Postscript	EP-S
Burroughs	AP9208	R4080
	AP9215	R4080
Canon	LBP 8II	EP-S
	LPB-8A1	EP
	LPB-8A2	EP
	LPB-20	EP
	LPB-200S	EP

TABLE 1 **Laser printer compatibility chart** (continued)

Canon (continued)	LPB-4	EP-L
	LPB-CXCAD	EP
	PC 3/5/51	A15/A30
	PC 6/6re/7	A15/A30
	PC 11	A15/A30
	PC 10/14/20	PC 10/25
	PC 24/25	PC 10/25
	PC 70 µfiche	MP N/P
	PC 80 µfiche	MP N/P
Cenegraphics	8707	R4080
Centronics	Page Printer #8	EP
CLR	FormWriter 2	EP
CMT	800	R4080
Compugraphic	CG 300-PS	EP
	EP 308	EP
Computer Language	FormWriter 2	EP
	FormWriter 2A	EP
	FormWriter 2X	EP
	FormWriter 4	Ricoh
	FormWriter 10	EP
	FormWriter 10D	EP
	Res FormWrtr4	R4080
	Res FormWrtr8	R4080
Concept Technologies	ConceptWriter	EP
	Laser 8	EP
Cordata Inc	Intellipress Prtr	EP
	LP 300	EP
	LP 300X	EP
Corona Data Systems	Corona LP-300	EP
	Corona LP-300X	EP
Corporate Data Sci.	CDS 2300	EP
.	CDS 4300	EP
CPT	LP-6	EP
	LP-8GS	EP
	LP-8S	EP
	LP-300	EP
	Pageprinter 1	R4080
	PS-8	EP
Data Card	Troy 308	R4080
Data General	4557	EP
	4558	EP

TABLE 1 **Laser printer compatibility chart** (continued)

Datacopy	LP	EP
Datageneral Corp.	DG Model 4557	EP
.	DG Model 4558	EP
Datapoint	7410 StarBeam	EP
DEC	LN 03 Plus	R4080
	LN 03	R4080
	Scriptprinter	R4080
Decision Data	6408	R4080
	6408-2	R4080
	6415 15PPM	R4080
	6415	R4080
Destiny Tech Corp	Laseract II	R6000
Detewe	8A1	EP
DTC/Kidron	Octave LP	EP
E.L.T.	Labelmaster 10	R4080
	Labelmaster 20	R4080
EFS	FormWriter 2	EP
	FormWriter 2X	EP
Epson	GQ-3500	R6000
	GS-3500	R6000?
Ericsson	7160	R6000
Facit	Opus 3 15PPM	R4080
	Opus P60...	R6000
	Opus1	R4080
	P7080 (Opus 2E)	R6000
	Opus 4	R6000
	P7080A	R4080
	P7150 (Opus 3)	R4080
Formwriter	II LaserPrinter	EP
GBT	6630	EP
	6630DW	EP
	6630LS	EP
	6633XP	EP
	6634XP	EP
	6635XP	EP
GDS Systems	Laser 5224	EP
Genecom	NewGen Laser	EP-S

TABLE 1 **Laser printer compatibility chart** (continued)

General Computer	Business LP	R6000
	Personal LP	R6000
	Personal Lsr PTI	R4080
Genesis Computer	LaserSet Ptr	EP
Genigraphics	8707 8PPM	R4080
	8707	Ricoh
Getronics	Visa LSR-6000	R6000
Graphic Softwr Syss	Concept Writer	EP
GTC Technologies	Blaser	EP
	Blaser Five	EP
	Blaser II	EP
	BlaserStar	EP
Hanzon	LP-3000	R4080
	LP-5000	R4080
Harris	H165	R4080
Harris/Lanier	LS-6	R6000
	LS-8	R4080
Hayes μCmptr Prdcts	LaserPrinter	EP
Hewlett Packard	1686TA	EP
	2686TA	EP
	2687A	R4080
	2688A	R4080
	DeskJet	Inkjet
	LaserJet	EP
	LaserJet 2686	EP
	LaserJet 500 Plus	EP
	LaserJet Plus	EP
	LaserJet PlusT	EP
	LaserJet II	EP-S
	LaserJet III	EP-S
	LaserJet IIP	EP-L
HISI	LP-601	R6000
Honeywell	Italia Laserpage	R4080
	Model 80	R4080
IBM	4216	R6000
	4216 Persnl PP	R6000
	Pagelaser 4216	R4080
	Persnl PP 2	R6000

TABLE 1 **Laser printer compatibility chart** (continued)

Imagen	8/300	EP
	12/300	R4080
	2308	EP
	3308 XP	EP
	ImageStation	EP
	Innovator	EP
Imprint Technologies	Light Writer	EP
Informer Compt	Terml 287-LP	EP
InformTech	PC-LPI	R6000
	LP-6000	R6000
Interkom	I-4908	EP
Interleaf	LPI-308	EP
	LPR-308	EP
	OPS-2000	EP
Itek	Digitek	EP
	Digitek Lsr	EP
	PTW Laser	EP
Kel	Kel M5300	EP
Kodak	EktaPrint 1308	EP
Koolshade	KS-LP28	R4080
Kyocera	1010/2010	Kyocera
	1000/1200	Kyocera
Laser Barcode Syst	Barcode Printer	EP
Laser Connection	PS Jet Plus	EP
	QMS-PS 810	EP-S
LaserData Inc	LaserView	R4080
Laserlink Sys	Jet Plus	EP
	Jet Twinmax	EP
LaserMaster	4081	R4080
	LTD	R4080
	Laser Max 1000	EP-S
	XT/RP	R4080
	XP/RP	R4080
LaserSoft	PT SYS II	R4080
Linotype	LaserPrinter 8/4	EP
Management	MGI Laser	EP

TABLE 1 **Laser printer compatibility chart** (continued)

Memorex	2108B	R4080
	2115	R4080
Mitek Systems	LaserShare 2105	EP
	LaserShare 2115	EP
	Model 100T	EP
	Model 110T	EP
	Systems 115T	EP
	Systems 120T	EP
	Systems 125T	EP
	Systems 2125	EP
Mnemos	6000	R4080
	SmartWrtr 80+	EP
	Laser 8-Big Kiss	EP
	Laser 8-PS800	EP
	Laser 8-PS800+	EP
NBI Inc	908	R4080
	IWS LaserWriter	EP
	LaserPrinter	EP
NCR	1510	R4080
	6406	EP
	6416	EP
	6436	R4080
NEC	LaserSmith 415	EP
	SilntWrtr LC890	NEC 890
	210	EP-S
	290	EP-S
	Model II 90	Minolta
Newgen Turbo	PS/300	EP-S
NeXT	Laser Printer	EP-S
North Atlantic	LaserII	EP
Oasys	850 LaserPrinter	EP
	LaserPro 805-C	EP
	LaserPro 805R	R4080
	LaserPro 810-C	EP
	LaserPro 812	R4080
	LaserPro 820-C	EP
	LaserPro 820-R	R4080
	LaserPro 1510	R4080
	LaserPro	EP
	Laserpro 810R	R4080
Octave Systems	LaserOctave	EP
Office Automation	LaserPro 8C	EP
	LaserPro 805-C	EP
	LaserPro 810-C	EP

TABLE 1 **Laser printer compatibility chart** (continued)

Okidata	Laserline 6 LaserLine 6 Plus	R6000 R6000
Olvetti	PG101	EP
Olympia	LaserLine 6 LaserStar 6	R6000 R6000
OMT	800	R6000
Packard Bell	PB-83PS PB-8300 PB-8300CP PB-9000 Plus	EP EP EP R6000
PCPI	DaiseyLaser Laserimage 1000	EP R6000
Personal Comp	Daisylaser 1000 DaisyLaser 2000 Laserimage Laserimage 2000 Laserimage 3000	EP R4080 R4080 R4080 R4080
Phillips Info System	LaserPrinter PLP-15	EP-S R4080
Prime Computer Inc	8PPM	EP
QMS	410 Big Kiss Big Kiss II ConceptWriter Kiss LaserGrafix 800 LasrGrafix 800 II LaserGrafix 1500 LaserGrafix 1510 PS-800 PS-800 II PS-800+ PS-810 PS-820 SmartScript 800 SmartWrtr 8/3X SmartWriter 80 SmartWriter 80+ Smartwriter 150	EP-L EP EP-S EP EP EP EP R4080 R4080 EP EP EP EP-S EP-S EP EP EP EP R4080
Quadram	Formsptr Plus Formsprinter Quadlaser DP8 Quadlaser FP8 Quadlaser Plus Quadlaser PS	R4080 R4080 R4080 R4080 R4080 R4080

TABLE 1 **Laser printer compatibility chart** (continued)

Quadram (continued)	Quadlaser PS+	R4080
	Quadlaser PP8	R4080
	Quadlaser WS8	R4080
Radio Shack	LP1000	R6000
Renful	CLP2000	R4080
	ELP2000	R4080
Ricoh	LP1060	R4080
	LP4080	R4080
	LP4080R	R4080
	LP4120	R4080
	LP4150	R4080
	LP40814I	R4080
	PC Laser 6000	R6000
	LaserLine 6000	R6000
	LP-1060	R6000
Rise Tech	EPT-1	EP
	ETP-1	EP
Sony	IPL 1340	EP
	OAP 5108	R4080
Star Micronics	LaserPrinter 8	EP-S
Steinic	L-2060	R6000
	LaserPrinter	EP
	Sun LaserWriter	EP
Syntrex	Synjet	R4080
Tab	Series 100	EP
Talaris	610/620	EP
	802	EP
	810/812	EP-S
	System T-160	EP
	System T-800	EP
	System T-1500	R4080
Talaris Systems	810/812	EP-S
T-1590	R4080	
Tandy/Radio Shack	LP-1000	R6000
Telex	Laserprinter	R4080
Telscan Systems	Barcode Reader	EP
Texas Instruments	OmniLaser 2015	R4080
	OmniLaser 2106	R6000
	OmniLaser 2108	R4080
	OmniLaser 2115	R4080

TABLE 1 **Laser printer compatibility chart** (continued)

TGV	Data Laser 1	R4080
	Data Laser 6	R6000
	Data Laser 8	R4080
Troy Data Corp	308	R4080
	BPPM	R4080
U.S. Lynx	Lynxlaser 8PPM	R4080
	Lynxlasr 15PPM	R4080
Unisys	AP 9208 Mod I	R4080
	AP 9215 Mod I	R4080
	AP 9215-1 Mod I	R4080
Varityper	LP2300	R4080
Wang	LCS-15	R4080
	LPS-8	EP

Index

More from Peachpit Press. . .

Art of Darkness (with disk)

Erfert Fenton

This book is the perfect companion to AFTER DARK, the world's most popular screen saver and one of the top-selling utility programs of all time. It explains in straightforward language how to install, operate, and customize the program and its follow-up product, MORE AFTER DARK, as well as how to create modules for new screen savers. Included are nine new AFTER DARK modules created exclusively for this book. *(128 pages)*

Desktop Publisher's Survival Kit

David Blatner

Desktop publishing can create fabulous looking documents, but often novice publishers find themselves stuck with pages that don't look quite as good as they expected. *The Desktop Publisher's Survival Kit* is a book/disk package that provides insights into common technical problems on the Macintosh: troubleshooting print jobs, working with color, scanning, and selecting fonts. It also discusses in detail everything from graphics file formats and digital fonts to word processing, color, typography, style sheets and printing techniques. A disk containing 12 top desktop publishing utilities, 400K of free clip art, and two fonts is included in the package. *(176 pages)*

Illustrator Illuminated

Clay Andres

This book is for people who want to know more about Adobe Illustrator than the manuals can tell them. *Illustrator Illuminated* uses full-color graphics to show how professional artists use Illustrator's tools to create a variety of styles and effects. Each chapter shows the creation of a specific illustration from concept through completion. Additionally, it covers using Illustrator in conjunction with Adobe Stream-line and Photoshop. *(200 pages)*

The Little Mac Book, 3rd Edition

Robin Williams

Praised by scores of magazines and user group newsletters, this concise, beautifully written book covers the basics of Macintosh operation. It provides useful reference information, including charts of typefaces, special characters, and keyboard shortcuts. Totally updated for System 7. *(336 pages)*

The Little Mac Word Book

Helmut Kobler

For users new to Microsoft Word who need to learn the program fast and for experienced users who want to familiarize themselves with the features of version 5.0, this book is just the ticket. In addition to discussing Word basics, *The Little Mac Word Book* provides concise and clear information about formatting text; using Word with Apple's new System 7 operating system; taking advantage of Word's writing tools, including its spelling checker, thesaurus and grammar checker; setting up complex tables; and much more! (240 *pages)*

The Little QuicKeys Book

Steve Roth and Don Sellers

This handy guide to CE Software's QuicKeys 2.0 explores the QuicKeys Keysets and the different libraries QuicKeys creates for each application; shows how to link together functions and extensions; and provides an abundance of useful macros. *(288 pages)*

The Macintosh Bible, 4th Edition

Arthur Naiman, Nancy E. Dunn, Susan McAllister, John Kadyk, and a cast of thousands

It's more than just a book—it's a phenomenon. Even Apple's own customer support staff uses it. Now the Fourth Edition is here, and its 1,248 pages are crammed with tips, tricks, and shortcuts to get the most out of your Mac. And to make sure the book doesn't get out-of-date, three 30-

page updates are included in the price (we mail them to you free of charge). Every Mac owner should have one. *(1,248 pages)*

The Macintosh Bible Guide to FileMaker Pro 2.0

Charles Rubin

Claris's FileMaker Pro product manager enthusiastically declared this book a "must for every FileMaker Pro user." Best-selling author Charles Rubin offers fast relief for FileMaker users of all levels, providing clear and understandable solutions for scores of the most common problems. *(464 pages)*

The Macintosh Bible "What Do I Do Now?" Book, 2nd Edition

Charles Rubin

Completely updated through System 7, this bestseller covers just about every sort of basic problem a Mac user can encounter. The book shows the error message exactly as it appears on screen, explains the problem (or problems) that can produce the message, and discusses what to do. This book is geared for beginners and experienced users alike. *(352 pages)*

The Mac is not a typewriter

Robin Williams

This elegant guide to typesetting on the Mac has drawn raves for its clearly presented information, friendly tone, and easy access. Quick and easy chapters cover the top twenty things you need to know to make your documents look clean and professional: em dashes, curly quotes, spaces and indents, white space, and visual balance. It's a primer that novices can pick up quickly, and even the pros can keep going back to, getting more every time. *(72 pages)*

Photoshop 2.5: Visual QuickStart Guide

Elaine Weinmann and Peter Lourekas

The author of our award-winning *QuarkXPress 3.1: Visual QuickStart Guide* does it again. This is an indispensable guide for Mac users who want to get started in Adobe Photoshop but who don't have a lot of time to read books. Covers how to use masks, filters, colors, tools, and much more. *(264 pages)*

The Photoshop Wow! Book (with disk)

Linnea Dayton and Jack Davis

"Tips, tricks, and techniques for getting the most out of Adobe Photoshop" is the subtitle of this book—and it fits! The book teaches by showing—using step-by-step illustrations and clear, practical guidelines to present the basic technical information that users need in order to use Adobe Photoshop effectively and profitably. The book is heavily illustrated in full color throughout and includes artwork created by top Photoshop graphic designers and artists, accompanied by brief explanations of what techniques were used. The disk contains special effects filters, textures, and other program enhancements. *(200 pages)*

The QuarkXPress Book, 3rd Edition:

David Blatner and Keith Stimely

This best-selling guide to the world's most powerful desktop publishing program is required reading for any serious XPress user. It explains how to import and modify graphics and provides tips for printing, using XTensions and other applications to customize the user interface. Additionally, it includes a section outlining XPress 3.1's numerous new features and techniques. *(640 pages)*

QuarkXPress 3.1: Visual QuickStart Guide (Mac Edition)

Elaine Weinmann

Winner of the 1992 Benjamin Franklin Award, this quick reference guide teaches QuarkXPress 3.1 with lots of illustrations and screen shots to make each feature of the program absolutely clear. It's a terrific way to get introduced to the program in just a couple of hours. The book begins with the basics, and then proceeds through the program's various features and capabilities. Lots of illustrations and screen shots make each feature of QuarkXPress absolutely clear. *(200 pages)*

QuarkXPress Tips & Tricks
David Blatner and Eric Taub

All the smartest, most useful shortcuts and techniques from Peachpit's bestselling *QuarkXPress Book* plus many more are packed into this book. *QuarkXPress Tips & Tricks* provides quick answers to common questions, as well as insights on techniques that will make you a power user. Each chapter covers a distinct aspect of QuarkXPress, such as document construction, type and typography, copy flow, color, and printing. An excellent companion for any QuarkXPress user. *(282 pages)*

Real World FreeHand 3
Olav Martin Kvern

The ultimate insider's guide to FreeHand, this authoritative and entertaining book gives pros an abundance of street-smart information, yet manages to make even complex issues accessible to novices. After laying out the basics, the book concentrates on advanced techniques. It delves into how to fine-tune FreeHand and your Macintosh, with tips on rewriting PPDs, working with other applications, printing, working with color, and more. *(512 pages)*

Silicon Mirage
Steve Aukstakalnis and David Blatner

Virtual reality is the amazing new technology of "pretend worlds," where individuals can completely immerse themselves in computer-generated environments. *Silicon Mirage* provides an easily understandable explanation of the "virtual senses" already possible, the strikingly broad array of fields where virtual reality is having an impact, and the breathtaking horizons yet to be discovered. *(300 pages)*

The Underground Guide to Laser Printers
Editorial Staff of The Flash

The *Underground Guide to Laser Printers* shows how to save hundreds of dollars in laser printer supplies, repairs, maintenance, and upgrades; how to get the best quality and productivity out of your printer; and how to push the limits of your printer with special techniques for creating negatives, overheads, separations, and iron-on transfers. It's a collection of the best articles from four years of The Flash, a popular hands-on newsletter for laser printer users. *(176 pages)*

Zen and the Art of Resource Editing (with disk)
Derrick Schneider et al.

This book introduces the beginner to ResEdit 2.1, one of the most useful tools ever designed for the Macintosh. It covers ResEdit from a nonprogrammer's point of view, and shows how to customize the Finder, menus, keyboard and icons. The book contains a disk with the lastest version of ResEdit and 1400K of sample resources. *(240 pages)*

Order Information:

How soon will I get my books?

UPS Ground orders arrive within 10 days on the West Coast and within three weeks on the East Coast. UPS Blue orders arrive within two working days anywhere in the U.S., provided we receive a fax or a phone call by 11 a.m. Pacific Time.

What about backorders?

Any book that is not available yet will be shipped separately when it is printed.
Requesting such books will not hold up your regular order.

What if I don't like it?

Since we're asking you to buy our books sight unseen, we back them with an *unconditional money-back guarantee*. Whether you're a first time or a repeat customer, we want you to be completely satisfied in all your dealings with Peachpit Press.

What about shipping to Canada and overseas?

Shipping to Canada and overseas is via air mail. Orders must be prepaid in U.S. dollars.

Order Form

To order, call:

(800) 283-9444 or (510) 548-4393 (M-F) • (510) 548-5991 fax

#	Title	Price	Total
	Art of Darkness (with disk)	19.95	
	Desktop Publisher's Survival Kit (with disk)	22.95	
	Illustrator Illuminated	24.95	
	The Little Mac Book, 3rd Edition	16.00	
	The Little Mac Word Book	15.95	
	The Little QuicKeys Book	18.95	
	The Macintosh Bible, 4th Edition	32.00	
	The Macintosh Bible Guide to FileMaker Pro 2.0	22.00	
	The Macintosh Bible "What Do I Do Now?" Book	15.00	
	The Mac is not a typewriter	9.95	
	Photoshop 2.5: Visual QuickStart Guide (Mac Edition)	18.00	
	The Photoshop Wow! Book	35.00	
	The QuarkXPress Book, 3rd Edition (Macintosh)	29.00	
	QuarkXPress 3.1: Visual QuickStart Guide (Mac Edition)	14.95	
	QuarkXPress Tips and Tricks	21.95	
	Real World FreeHand	27.95	
	Silicon Mirage	15.00	
	The Underground Guide to Laser Printers	12.00	
	Zen and the Art of Resource Editing	24.95	

SHIPPING:	First Item	Each Additional
UPS Ground	$4	$1
UPS Blue	$7	$2
Canada	$6	$4
Overseas	$14	$14

Subtotal		
8.25% Tax (CA only)		
Shipping		
TOTAL		

Name

Company

Address

City State Zip

Phone Fax

❑ Check enclosed ❑ Visa ❑ MasterCard

Company purchase order #

Credit card # Expiration Date

Peachpit Press, Inc. • 2414 Sixth Street • Berkeley, CA • 94710

Your satisfaction guaranteed or your money will be cheerfully refunded!